Exploring the Night Sky with Binoculars

On a clear starry night, the jewelled beauty and unimaginable immensity of our universe is awe-inspiring. Stargazing with binoculars is rewarding and may begin a lifelong hobby! Patrick Moore has painstakingly researched *Exploring the Night Sky with Binoculars* to describe how to use binoculars for astronomical observation. He explains basic astronomy and the selection of binoculars, then discusses the stars, clusters, nebulæ and galaxies that await the observer. The sky seen from the northern and southern hemispheres is charted season by season, with detailed maps of all the constellations. The reader can also observe the Sun, Moon, planets, comets and meteors. With many beautiful illustrations, this handbook will be helpful and encouraging to casual observers and those cultivating a more serious interest. The enjoyment of amateur astronomy is now available to everybody.

PATRICK MOORE, host of the monthly television show *The Sky at Night* since 1957, is a distinguished and prolific author of more than 70 astronomy books. He has served as President of the British Astronomical Association. His contributions to the public understanding of astronomy have been marked by special awards from the Royal Astronomical Society, the British Astronomical Association, the Astronomical Society of the Pacific and the Italian Astronomical Society. Patrick Moore is a great enthusiast, always encouraging anyone with an interest in the night sky to get out and observe.

Exploring the

NIGHT SKY

with Binoculars

FOURTH EDITION

Patrick Moore

CAMBRIDGE
UNIVERSITY PRESS

PUBLISHED BY THE PRESS SYNDICATE OF THE UNIVERSITY OF CAMBRIDGE
The Pitt Building, Trumpington Street, Cambridge, United Kingdom

CAMBRIDGE UNIVERSITY PRESS
The Edinburgh Building, Cambridge CB2 2RU, UK
40 West 20th Street, New York, NY 10011–4211, USA
10 Stamford Road, Oakleigh, VIC 3166, Australia
Ruiz de Alarcón 13, 28014 Madrid, Spain
Dock House, The Waterfront, Cape Town 8001, South Africa

http://www.cambridge.org

First published 1986
First paperback edition 1988
Reprinted 1989, 1990, 1993
Third edition 1996
Reprinted 1997, 1998, 1999
Fourth edition 2000

Printed in the United Kingdom at the University Press, Cambridge

Typeface Palatino 10/12 pt   *System* QuarkXPress® [DS]

A catalogue record for this book is available from the British Library

Library of Congress Cataloguing in Publication data

Moore, Patrick.
Exploring the night sky with binoculars/Patrick Moore.– 4th ed.
    p.   cm.
ISBN 0 521 79053 0 – ISBN 0 521 79390 4 (pbk)
1. Astronomy–Observers' manuals. I Title.
QB63.M63 2000
520'.22'3–dc21   00-028913

ISBN 0 521 79053 0 hardback
ISBN 0 521 79390 4 paperback

Contents

3253

Preface to the Fourth Edition

A few further amendments have been made in order to bring the text fully up to date.
I am very grateful to Dr Alice Houston for her help.

Selsey, May 2000   PATRICK MOORE

1

The Night Sky

A few weeks ago I went out in the late evening, armed with a pair of 7×50 binoculars. The sky was almost dark, and beautifully clear. Above me the stars shone down, and when I turned my binoculars toward the lovely cluster of the Seven Sisters I realized, yet again, that to take a real interest in astronomy there is no need to have a powerful telescope.

My own interest in astronomy dates from the age of six, when I picked up a book which happened to be lying around, and began to read it. I was fascinated, and took what I still believe to be the correct steps. I made sure that I could understand the basic facts, and then I equipped myself with an outline star map and started to learn my way around the sky. It did not take nearly so long as I had expected. There are less than three thousand stars visible with the naked eye at any one time, and the constellation patterns never change, so that once you have identified a group there is no real problem in recognizing it again. I found the Great Bear, the Little Bear, the Swan and the Square of Pegasus; when winter came I used Orion, the Hunter, as a guide to other constellations, and before long I could identify the brightest stars with no trouble at all. I could see, too, that they were not all alike. Some were white, while others were distinctly orange-red, and a few were bluish – notably Vega, which is almost overhead from England during summer evenings. Then there were the planets, which looked like stars but which I tracked as they wandered slowly around from one constellation to another. I remember that Venus and Jupiter were both on view, and there was no mistaking them, because they were so brilliant. Next came Mars, identifiable because of its fiery red tint. Saturn was more of a problem; it looked to me like an ordinary star, and it puzzled me for some time because I could not find it on my outline map. I looked at the Moon, and made out the dark patches which we still call 'seas' even though there has never been any water in them. I watched for meteors or shooting-stars dashing across the sky, and within those first few months I was lucky enough to see a display of the Northern Lights – something which is not common from a latitude as far south as that of Sussex where I lived. The more I saw and read, the more my interest grew.

I wanted to make a real hobby out of astronomy. That, I thought, meant obtaining a telescope. I saved up my pocket money, Christmas presents, birthday presents and everything else until I had accumulated the princely sum of £7.10s ($12). That was enough to buy a proper telescope – a refractor

1

with an object-glass three inches (76 millimetres) in diameter. I still have that telescope, and I still use it. With it I learned my way around the Moon, saw the belts and satellites of Jupiter, the rings of Saturn and almost countless star-fields and clusters. I was 'hooked'.

But this was long ago; when I bought my telescope, which must have been in 1933, prices were very different from those of today. Look for a 75-mm refractor now, and you will have to pay something of the order of £600 ($1000), which is more than many people care to spend upon a hobby. Therefore, what often happens is that the would-be enthusiast gives up, and lets his interest in astronomy fade into the background.

Binoculars provide a satisfactory answer. True, they are of low magnification when compared with telescopes, but they will give immense pleasure, and then can even be used for a certain amount of real scientific work; after all, astronomy is still just about the only science in which the amateur can play a useful rôle. There are some specialist observers who use binoculars and nothing else. The cost is comparatively modest. You can probably pick up a reasonable pair of binoculars for £50 ($75) or so, and with £100 ($150) you have a wide choice. Also, most households can muster binoculars of some sort, probably used originally for bird-watching or looking at ships out to sea. Even old-fashioned opera-glasses are not to be despised.

The value of binoculars in astronomy is not widely appreciated, and this book is an attempt to help in putting the record straight. I will not pretend for a moment that binoculars can rival telescopes, but they have their own particular advantages, quite apart from the fact that they can be bought comparatively cheaply.

Before setting out upon a tour of the sky, it is, I think, worth spending a few pages in giving an outline of the basic facts. Most people will know them already (in which case, you have my full permission to skip the rest of this chapter and proceed straight to page 8), but it is as well to be sure. I still come across people who confuse the science of astronomy with the ancient pseudo-science of astrology, which links the movements of the planets with human character and destiny. My comment here is that astrology proves one thing only: 'There's one born every minute!'

The Earth upon which we live is a planet, moving round the Sun at a distance of 150 000 000 kilometres.* This may sound a long way, but it is not much to an astronomer, who has to become used to dealing with vast distances and immense periods of time. The Sun is a normal star, not nearly so luminous as many of those visible on any clear night. It appears so glorious in our skies simply because, cosmically speaking, it is upon our doorstep. Represent the Earth–Sun distance by 25 millimetres, and the nearest star will be over 6½ kilometres away. The Sun is the centre of the Solar System, which is made up of nine planets (including the Earth), the moons or satellites of the planets, and various bodies of lesser importance such as comets and meteoroids. On the cosmical scale the Solar System is local, and it is the only part of the universe which we can hope to explore physically, at least for the moment.

---

* A Metric–Imperial conversion table may be found on p. 206.

The Sun itself is a globe of gas, large enough to swallow up more than a million bodies the volume of the Earth, and fiercely hot even at its surface – which is why nobody should ever look directly at it through any kind of optical instrument, a point to which I will return later. It is not 'burning' in the conventional sense of the term, but is more in the nature of a huge, controlled atomic bomb, inasmuch as it is producing its energy by nuclear reactions taking place deep inside it. We depend entirely upon the Sun; without it the Earth and the other planets would never have been formed.

The ordinary stars are of the same kind, though of course they differ in detail. They are so far away that they appear simply as points of light, and no telescope will show them as disks. Also, they keep to virtually the same relative positions in the sky, so that the constellation patterns are the same now as they must have been in the time of George Washington, William the Conqueror or Julius Cæsar. The stars are not genuinely fixed in space, and are moving around in all sorts of directions at all sorts of speeds, but their remoteness slows down their apparent individual movements so much that the naked-eye observer will not notice them. If you compare the speed of a bird flying at tree-top height with that of a jet-aircraft high above you, the bird will seem to move quickly while the jet will crawl even though the jet is really much the faster of the two. The rule, is 'The further, the slower', and the stars are so distant that they seem to be fixed. Indeed, they were once commonly known as 'fixed stars' to distinguish them from the wandering stars, or planets.

We all know that the stars seem to travel across the sky from east to west, but this is due entirely to the fact that we live upon a spinning globe. The Earth rotates from west to east, and so the entire sky appears to revolve, carrying the Sun, Moon, stars and planets with it. The axis of rotation points northward to the north celestial pole, marked fairly closely by the brightish star known as Polaris; the south celestial pole is not near any bright star which southern-hemisphere sailors had cause to regret in the days when navigation was very much a hit-or-miss affair. And just as the Earth's equator divides the world into two hemispheres, so the celestial equator divides the sky into two halves.

Let us look at this situation a little more closely. The Earth's axis is tilted in its path or orbit by an angle of 23½ degrees to the perpendicular, which explains the seasons; when the North Pole is tilted toward the Sun (position 1 in the diagram) the northern hemisphere receives the full benefit of the solar radiation, while when the North Pole is tilted in the opposite direction

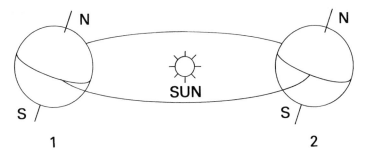

(position 2) it is the southern hemisphere which is favoured. Actually, the Earth's orbit is not quite circular; our distance from the Sun ranges from 147 000 000 kilometres in December out to 152 000 000 kilometres in June, but this is not enough to make much difference, and in any case the effect is compensated for by the fact that there is much more ocean in the southern hemisphere, tending to stabilize the temperature. Water heats up more slowly than land, but is better at retaining its warmth.

Go to the North Pole, and you will find that Polaris is overhead (or almost so; it is less than one degree of arc away from the polar point). The celestial equator will lie all round the horizon, and the stars will move in circles so that the stars north of the equator are always on view when the sky is dark enough, while those south of the equator can never be seen at all. As you move further south, Polaris will drop in altitude; from London, for example, it is only a little more than 50 degrees above the horizon, and some of the southern-hemisphere stars have come into view. From the equator, the two celestial poles lie on opposite horizons; from countries such as Australia, Polaris never rises and the south celestial pole never sets. From the South Pole the situation is exactly opposite to that at the North Pole, with the southern-hemisphere stars permanently visible except when they are blotted out by the brightness of the sky.

At an early stage, thousands of years ago, the stars were divided up into definite constellations. For instance, most Europeans can recognize the Great Bear, while Australians and South Africans are equally familiar with the Southern Cross. Yet the constellation patterns really mean nothing at all, because the stars are at very different distances from us, and the stars in any one particular constellation are not genuinely associated with each other; they merely happen to lie in much the same direction as seen from Earth. We happen to follow the constellation patterns drawn up by the Greeks, well before the time of Christ. If we had chosen, say, the Chinese or the Egyptian system, our star-map would look quite different – even though the stars themselves would be in exactly the same positions.

What about distances? With the stars, the mile or the kilometre is much too short a unit to be convenient, just as it would be clumsy to measure the distance between London and New York in millimetres. Luckily there is a better unit available. Light does not travel instantaneously; it flashes along at 300 000 kilometres per second, and in a year it covers 9.46 million million kilometres. This is the 'light-year', which, please note, is a measure of distance and not of time. It takes only about 8½ minutes for light to reach us from the Sun, but over four years from the nearest of the fixed stars. This also means that we see the universe not as it is now but as it used to be in the past. Deneb, a bright star in the constellation of the Swan, is about 1600 light-years away, so that we see it today as it used to be when the Romans were making ready to evacuate Britain.

The star-system of which our Sun is a member is known as the Galaxy. It contains about a hundred thousand million stars, and it is flattened; I have compared its shape, rather unromantically, with that of two fried eggs clapped together back to back. When we look along the main plane of the system we see many stars in almost the same line of sight, which explains the lovely band of the Milky Way. Appearances can be deceptive. The stars in the Milky Way, easily seen with binoculars, may look as though they are in danger of crashing into each other, but they are not genuinely crowded, and direct collisions must be very rare indeed.

Our Galaxy is not the only one. There are many millions of others, some of them much larger than ours. Three of these external systems are clearly visible with the naked eye (one in the northern hemisphere and two in the far south of the sky) and binoculars will show several more, though they are so remote that photographs taken with large telescopes are needed to bring out their details. By now we can range out to immense distances. Whether the universe is finite or infinite is something which we do not yet know, and it is not a problem which need concern us here, fascinating though it undoubtedly is.

Now let us come much nearer home and look at the Solar System, our own particular part of the universe. First there are the planets, of which nine are known. Mercury, Venus, the Earth and Mars make up the inner group; then comes a wide gap, in which move thousands of midget worlds known as asteroids, and then we come to the four giants, Jupiter, Saturn, Uranus and Neptune. The planetary system is completed by Pluto, a far-away, peculiar world which seems to be in a class of its own.

The planets move round the Sun at different distances in different times. Our own revolution period or 'year' is, of course, 365¼ days; that of Mercury is a mere 88 Earth-days, while Neptune takes over 164 years to complete one circuit. Most of their orbits are not very different from circles, though those of Mercury, Mars and Pluto are considerably more elliptical than that of the Earth. The planets have no light of their own, and shine because they are illuminated by the Sun. This tends to make us think that they are more important than they really are. Venus, Jupiter, and Mars at its best, are much more brilliant than any star, while Saturn and Mercury are bright enough to be noticeable; all these five were known in ancient times, though the three outer planets were discovered only in the telescopic era – Uranus in 1781, Neptune in 1846 and Pluto as recently as 1930. Only Pluto is too faint to be seen with binoculars.

Because the planets are so much closer than the stars, they move about against the starry background, though they keep within certain well-defined limits. This was how they were originally distinguished from the stars. The ancient Greeks worked out their movements with surprising accuracy, and were able to predict how they would behave.

All the planets except Mercury and Venus are attended by satellites. We have one natural satellite, our familiar Moon; Saturn has at least 18, though most of them are fairly small.

The Moon is our companion in space. It keeps together with us as we journey round the Sun, and moves at a mean distance of only 384 400 kilometres from us, so that anyone who flies ten times round the Earth will cover a distance as great as that between the Earth and the Moon. This, of course, is why the Moon dominates the night sky. It is a small world, with a diameter of only 3476 kilometres as against 12 756 kilometres for the Earth; represent the Earth by a tennis-ball, and the Moon will he about the size of a table-tennis ball, Its orbital period is just over 27 days, or a little less than a calendar month.

5

Because the Moon shines by reflected sunlight, only half of it can be lit at any one time, and the regular phases depend upon how much of the 'day' side is turned in our direction. When the Moon is almost between the Earth and the Sun, its dark or night side faces us, and the Moon is new, so that it cannot be seen at all. On the far side, the illuminated hemisphere is turned toward us, and the Moon is full. At other times we see a crescent, a half, or a three-quarter (gibbous) shape. When the Moon is near full it drowns all but the brighter stars, so that star-gazing is best done when the Moon is absent from the night sky.

Comets have been termed the stray members of the Solar System. They are not massive and substantial, as the planets are; a comet is made up of thin gas together with icy particles and what may be termed 'dust'. In most cases their paths are very elliptical, and since they too depend upon reflected sunlight we can see them only when they are in the inner part of the Solar System. The only bright comet which we see regularly is Halley's, named in honour of the astronomer who first worked out the way in which it moves. Halley's Comet returns every 76 years; it was bright in 1910, after which it moved back into the outer part of the system until its latest return in 1986. All other brilliant comets take so long to complete one circuit of the Sun that we never know when or where to expect them, and they are always apt to take us by surprise.

Though the twentieth century was rather comet-barren, there were two spectacular visitors in the closing years. The first was the comet of 1996, discovered by Japanese amateur Y. Hyakutake. It was a naked-eye object from March to May, and was probably the loveliest comet I have ever seen; it was delicate green in colour, with a long slender tail. It was also in the far north of the sky when at its best, and binoculars showed it magnificently. Unfortunately it was not bright for long – and it was in fact a small comet, but it came within about 10 000 000 miles of us, which by cometary standards was very close indeed.

It was outshone by its successor, Hale–Bopp, which could almost be classed as 'great', and which remained a naked-eye object for over a year. It was probably the most-observed comet in history. There was a gleaming head with spiral structure, and a dust-tail as well as a gas-tail; the dust-tail was beautifully curved. Between July 1996 and October 1997 it was on view, and binoculars were ideal for observing it and following its changes. The nucleus became as bright as Arcturus, and the comet was a familiar object in the evening sky. It was in fact a very large comet, but it never came within 120 000 000 miles of us; if it had been as close as Hyakutake, it would have cast shadows. Everyone was sorry to see it depart, and it was still a reasonably bright telescopic object at the start of 2000. It will be back in just over 2000 years – but Hyakutake will not return to perihelion for about 15 000 years. Whether the twenty-first century will provide any great comets remains to be seen; we can only hope.

There are many faint short-period comets, some of which can be seen with binoculars. Moreover, binocular-owners have often discovered comets which were not previously known.

If you see an object moving noticeably across the sky, it cannot be a comet, which is millions of miles away; you have to watch a comet for hours to notice any relative movement at all. A quickly-shifting object may well be one of the artificial satellites which have been launched in large numbers since October 1957, when the Russians opened the Space Age by launching their football-sized space-craft Sputnik 1. But if the movement is very rapid, and the object vanishes after a second or two, it will be a meteor. A meteor is a tiny particle, usually smaller than a pin's head, moving round the Sun; if it dashes into the upper part of the Earth's atmosphere, it rubs against the air-particles and is so heated by friction that it perishes in the streak of luminosity which we call a shooting-star. Meteors are common enough, particularly during the early part of August in each year, when our Earth ploughs through a swarm of meteors and collects a large number of shooting-stars.

I have had to cram this opening chapter with facts; I hope that you have not found it indigestible. Now, having cleared the air, so to speak, let us turn to our main theme – binoculars.

2

Binoculars of Many Kinds

It is seldom that a week passes by without my having several letters on the same theme. 'I have become interested in astronomy, so I want to buy a telescope. I can spend up to £50 ($75) or so. What sort of telescope should I get?'

Writing back is always something I find depressing, because the plain, unpalatable fact is that telescopes today are expensive items. It used to be possible to obtain second-hand telescopes at low cost, but by now a really good, cheap second-hand telescope is about as common as a great auk. In my view it is rather pointless to spend much money on any refracting telescope with an aperture of less than 3 inches (76 mm) or a reflector with a main mirror less than 6 inches (152 mm) in diameter. I emphasize 'in my view' because not everyone will agree, and of course a very small telescope is a great deal better than nothing at all. But given a choice between, say, a 60-mm telescope and a pair of good binoculars, I would unhesitatingly prefer the binoculars.

Telescopes are of two main kinds. With the refractor, the light from the object to be observed is collected by an object-glass; the light is brought to focus, and the image is then enlarged by a second lens termed the eyepiece. Note that the function of the object-glass is to collect the light, and all the actual magnification is done by the eyepiece. Changing the eyepiece means changing the magnification. In general, it is fair to say that one can use a magnification of × 50 for each inch of aperture (forgive my temporarily reverting to Imperial measure!), so that a 3-inch refractor will bear a magnification of × 150, a 6-inch will bear × 300, and so on. There are times when this limit can be exceeded, but it is a good enough rule. This is why I always distrust advertisements which offer, say, a 60-mm telescope '× 300'. For one thing, it is misleading to quote any definite power for a telescope, because it is the eyepiece which determines the magnification. Secondly, anyone who believes that a 60-mm refractor will bear a power of × 300 is going to be bitterly disappointed.

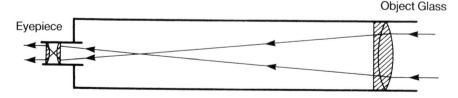

Light path in a refractor.

The answer is straightforward enough. Eyepieces are interchangeable (theoretically, at least), and so one can use any eyepiece with any telescope. In my observatory I have a 39-cm reflector, with which I can often use a power of × 600. If I used that particular eyepiece on my 76-mm refractor, I would still obtain a magnification of around × 600 – but the image would be so faint that it would be completely useless. The larger the telescope, the more light it can collect, and the higher the magnification which can be employed. Not that magnification is all-important; it is far better to have a smaller, well-defined image than a larger, blurred one even for observing the Moon or a planet.

The second type of telescope is the reflector. There are various optical systems, but the most common, at least in amateur hands, is the Newtonian, so named because the principle was first demonstrated by Isaac Newton more than three hundred years ago. Here, there is no object-glass. The light travels down an open tube until it hits a curved mirror at the bottom; the rays are then sent back up the tube on to a smaller, flat mirror placed at an angle of 45 degrees, so that the rays are directed into the side of the tube, where an image is formed and magnified by an eyepiece. With a Newtonian reflector, therefore, the observer looks into the tube rather than up it, so that for pointing to a planet or a star it is usually helpful to have a small refracting telescope mounted on to the side of the tube to act as a finder. (My 39-cm reflector has five finders, on the theory that the finder you want to use is always the one which you can't get at.)

Aperture for aperture, a refractor is more efficient than a reflector, and it is also easier to maintain. In fact it needs virtually no maintenance at all unless roughly treated, whereas the mirrors of a reflector have to be periodically re-coated with a thin layer of silver or aluminium. On the other hand a refractor is more costly, and in some ways less convenient to use.

If you set out to buy a telescope, take great care. If the object-glass of a refractor or the mirror of a reflector is of poor quality, the images will also be poor – and a bad telescope does not always betray itself at a glance. Therefore, I recommend either giving the instrument a thorough test, or else asking for an opinion by an optical expert. Also, pay attention to the mounting. The essential need is firmness. If the telescope is mounted upon a spidery stand, it will quiver charmingly in the slightest breeze, and the image will dance about so violently that it will be useless.

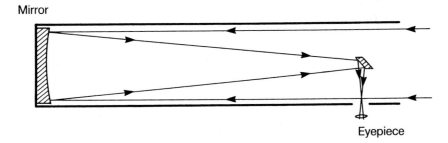

Mirror

Eyepiece

Light path in a Newtonian reflector.

The cost of a 152-mm reflector is much the same as that of a 76-mm refractor. There are pros and cons, and everything really depends upon the main interests of the observer; for example, anyone who wants to make regular studies of the Sun will be far better off with a refractor, while the deep-sky enthusiast will in general prefer a reflector. But again I stress that light-grasp is all-important, and I am not at all happy about the very small telescopes which can be bought cheaply. They will not be satisfactory, though they will show some pretty sights. I would recommend them only for the casual observer who wants little apart from views of lunar craters. Note too that most astronomical telescopes give an upside-down or inverted image. In fact all ordinary telescopes would do so but for the addition of an extra lens-system to turn the image the right way up again. Each time a ray of light passes through a lens it is slightly weakened. This does not in the least matter when looking at birds or distant ships, but it is not helpful when observing a planet or a star, when it is important to collect as much light as possible. Therefore, the extra lenses are left out. If you want an 'all-purpose' telescope, you will have to buy a terrestrial converter into which the eyepiece can be fitted.

A pair of binoculars is nothing more than two small refractors joined together. The main advantage is that the observer can use both eyes instead of only one. The field of view will be much wider than that of a small astronomical telescope, and the binoculars are handier, so that they can be carried around very easily. A word of warning here. Every time you use binoculars, put the safety-cord around your neck. If you drop the binoculars, which is only too easy, there will be no harm done, but if you have forgotten to use the cord disaster may result. Dropping a pair of binoculars several feet on to hard ground is emphatically not to be recommended.

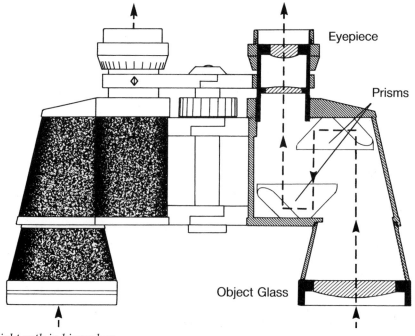

Light path in binoculars.

Binoculars are classified according to their magnifications and the diameters of their main object-glasses. Thus 7×50 indicates a magnification of 7, with each object-glass 50 millimetres across; 12×50 gives a magnification of 12, again with 50-mm object-glasses, and so on. And at the outset, it is very important to make a careful decision about just what type to buy.

As with telescopes, the larger the aperture the greater the light-grasp. but there are hazards too. With increased power, the field of view becomes smaller, and with increased aperture the binoculars become heavier. If you are going to buy only one pair, I would recommend something around 7×50. The magnification is adequate, the field pleasingly large, and the binoculars light enough to be hand-held without awkward shake. I have in my possession several pairs, and I think it will be useful to describe each of them, though naturally there is a whole range of sizes.

3×20. I obtained these during a visit to Tokyo some years ago! They are very light, very portable and decidedly useful, though the low magnification is a severe limitation, and unless you can find a pair very cheaply I would advise against them.

7×50. An excellent choice for general viewing. They will show considerable detail on the Moon, for example, and almost endless superb star-fields. For studying some kinds of objects, such as loose star-clusters, they are ideal because of their relatively large field.

8.5×50. Also very suitable. The light-grasp is adequate for this magnification, and the binoculars can still be comfortably hand-held. The increased power means that more details can be seen, but the field of view is noticeably reduced. I also have a pair of 8×56 binoculars, which are excellent.

11×80. These are very fine, with a reasonably large field and a good light-grasp, though they are rather too heavy to be properly hand-held.

12×40. Here the smaller aperture and higher magnification means that the images are not so bright as in the former pairs, and the field is smaller still. On the other hand, the lesser light-grasp is not really important for most kinds of viewing. The real trouble is that the more restricted field causes problems inasmuch as the binoculars have to be really steadily held. My usual method is to sit down and jam my elbows into my body. This works fairly well, but many people will consider that the time has come to consider a more rigid mounting.

20×70. Binoculars of this aperture cost a great deal of money. They are extremely difficult to hand-hold, which means that a mounting must always be carried around, and it should be rather firmer than a simple attachment to a camera tripod, because the binoculars are inevitably heavy. Yet when they have been properly set up, they can give splendid results. The Moon shows a wealth of detail, and the phases of Venus are easy to pick out, while striking views can be obtained of stellar objects. On the adverse side, the field has become too small for comfort , and it is not easy to line up the binoculars on the target. I would not suggest obtaining anything as powerful as $\times 20$ unless the aim is to undertake some really serious and specialized kind of observation.

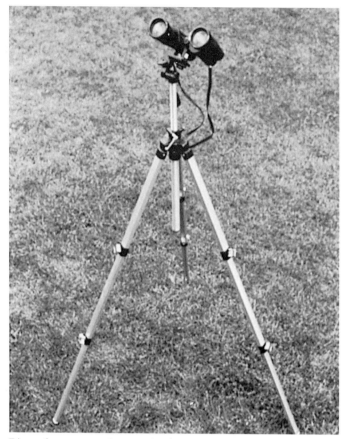

Binoculars mounted on a tripod.

Most photographic shops sell binocular mounts which can be screwed into the top of a camera tripod. This is satisfactory enough, though it is often necessary to perch the tripod on top of a platform or a solid table, as otherwise you will have to be something of a contortionist when looking at an object high in the sky (incidentally, the same is true of a small refracting telescope).

Making a mount for a pair of binoculars is a fairly simple problem of carpentry. Anyone who is reasonably skilful can knock up a mounting without difficulty, and it is possible even for someone who is as clumsy as I am. Again the main essential is steadiness. What you really want is the facility of moving the binoculars freely in any direction, in altitude and also in azimuth. With a magnification higher than × 12, a mounting of some sort is highly desirable. Recently it has become possible to obtain binocular mountings which take the form of a harness attached to the observer's chest. They are certainly effective, and make using high-power binoculars much easier. The general principle is shown in the photograph on p. 14, and with a little ingenuity it should be possible to devise something of the sort.

Larger and more ambitious binoculars can be found occasionally. George Alcock, one of Britain's most famous amateur astronomers, uses accurately-mounted binoculars with object-glasses as much as 150 millimetres across.

Binoculars mounted on a clamp which can easily be purchased.

This naturally gives him a tremendous light-grasp, but his main work is in hunting for new comets and new exploding stars or novæ, in which he has been remarkably successful (his present total is five comets and four novæ), and there are not many people who will attempt anything like this. For one thing, it needs immense patience. George Alcock knows the sky so well that he can identify 30 000 stars by memory, and can identify any newcomer at a glance. I well remember the occasion when he discovered a brightish nova, just about visible with the naked eye and easily seen with binoculars. He telephoned me at three o'clock in the morning, and asked me to confirm it. Using my 7 × 50 binoculars I was able to locate it at once, though I would not have had the slightest hope of finding it had I not been told exactly where it was.

In short: if you are going to have only one pair of binoculars, choose a magnification of between × 7 and × 10, with an aperture of between 40 and 60 millimetres. If you can afford several pairs, then I think that 7 × 50, 8.5 × 50 and 12 × 40 is a reasonable selection, though recently I borrowed a 10 × 60 pair and found it very useful.

The next question is: Where can they be bought? Photographic shops can supply them, and, unlike telescopes, it is possible to test them at once. Most binoculars are adequate in quality, though make sure that it is possible to

Binocular chest mounting.

bring both the twins into focus at the same time; if there is a double image, or if one of the twins will not focus sharply, beware. I have come across many people who have bought their binoculars from railway lost-property offices, though naturally one has to be doubly careful.

One final warning is worth repeating here. If you set out to buy a pair of binoculars, and decide to test them, look at the Moon by all means, but never, on any account, point them anywhere near the Sun. If you do, you run the risk of permanent blindness. I will return to this point later on, but I do not apologize for emphasizing it, because tragedies have occurred in the past.

I also advise against buying a pair of binoculars (or, for that matter, a telescope) by mail order. Always test your would-be purchase first.

3

Among the Stars

The last great astronomer of Classical times was Ptolemy of Alexandria, who flourished from around AD 120 to 180. It was he who wrote the famous book which has come down to us by its Arab translation, and is known as the Almagest. In it, Ptolemy summarized much of the scientific knowledge which had been amassed over the previous centuries. We owe a great deal to Ptolemy; without him, we would know much less about ancient science than we actually do.

Ptolemy compiled a star catalogue in which be listed 48 constellations. Most of them were named after mythological characters, though there were also a few everyday objects such as a Triangle and an Altar. All of Ptolemy's original groups are still to be found on our maps, though their boundaries have been modified in many cases.

So far as we know, Ptolemy spent his whole life in Alexandria, which is well north of the Earth's equator, so that he could not see the stars of the far south. Moreover, he had not covered the whole of the sky accessible to him. Later astronomers made extensive alterations, and added new groups made up of stars stolen from the existing 48. When the southernmost stars also were divided up, the whole situation became somewhat chaotic; various astronomers invented their own constellations, some of which were so small and so obscure as to be unworthy of separate identity. Moreover some of the names were curious – typical examples being Sceptrum Brandenburgicum (the Sceptre of Brandenburg) and Lochium Funis (the Log Line). Now and then there were proposals for reforming the entire system, and it was even suggested that the 12 constellations of the Zodiac might be named after the 12 Apostles. Mercifully ideas of this kind never met with much support, and finally, in 1932, the controlling body of world astronomy, the International Astronomical Union, lost patience. The total number of accepted constellations was reduced to 88, and the most cumbersome names were shortened; thus Apparatus Sculptoris, the Sculptor's Apparatus, became simply 'Sculptor'. It cannot be said that the result was entirely logical, and one is tempted to agree with a famous last-century astronomer, Sir John Herschel, that the constellations seem to have been drawn up so as to cause as much inconvenience as possible, but the system has become so well established that it is unlikely to be altered now. I think everyone would be sad to lose such familiar figures as the Great Bear, the hunter Orion, the Centaur and the Southern Cross.

The most brilliant stars were given individual names, most of which are Arabic (though a few are Greek: one example is Sirius in the Great Dog, the brightest star in the sky). Actually there are many stars with proper names, but in general these are used only for the first 30 or so, plus a few fainter stars of special interest – such as Polaris, Mizar in the Great Bear, and the variable star Mira, in Cetus, the Whale.

In 1603 a new star-map was compiled by a German amateur astronomer, Johann Bayer. He followed Ptolemy's constellations, adding a dozen of his own which are still accepted. More importantly, he introduced the system of allotting Greek letters to the stars in each constellation, from Alpha, the first letter of the Greek alphabet, through to Omega, the last. Since I will be using these letters in the present book, it will be useful to list them here. They are:

α Alpha	ε Epsilon	ι Iota	ν Nu	ρ Rho	φ Phi
β Beta	ζ Zeta	κ Kappa	ξ Xi	σ Sigma	χ Chi
γ Gamma	η Eta	λ Lambda	o Omicron	τ Tau	ψ Psi
δ Delta	θ Theta	μ Mu	π Pi	υ Upsilon	ω Omega

Thus the brightest star in Ursa Major, the Great Bear, should be Alpha Ursæ Majoris; the second brightest, Beta Ursæ Majoris, and so on. Inevitably the system has not been rigorously followed; thus in the Great Bear the brightest star is not Alpha, but Epsilon, which really ought to be fifth, while in Orion, Rigel (Beta Orionis) is brighter than Betelgeux (Alpha). In Sagittarius (the Archer) the three brightest stars are Epsilon, Sigma and Zeta. While writing this chapter, it occurred to me to look through the list of constellations to see in how many the five leading stars were, in order, Alpha. Beta, Gamma, Delta and Epsilon. There are only three: the Southern Cross, the Southern Triangle and Lupus.

Next we must define 'magnitude', which is a measure of a star's apparent brilliancy. The scale works in the same way as a golfer's handicap. The brightest performers have the lowest magnitudes; thus Aldebaran in Taurus, magnitude 1, is brighter than Polaris which is of magnitude 2. The faintest stars normally visible with the naked eye are of magnitude 6, and binoculars will go down to at least 8; my 20 × 70 pair will reach 9. At the other end of the scale we have stars which are above magnitude 1; Vega in Lyra (the Lyre) is almost exactly 0, while Sirius is –1.4. On the same scale, the Sun's magnitude is about –26. With the naked eye, or with binoculars, it is possible to distinguish between two stars which differ by only a tenth of a magnitude.

I have already commented that colours will not show up unless the intensity of the light is sufficiently great. Thus Betelgeux in Orion, one of the brightest stars in the sky, is clearly orange-red, but Mu Cephei, in the far north, shows almost no colour at all until you look at it with binoculars, when it gives the impression of a glowing coal, and earns its nickname of 'the Garnet Star'. Mu Cephei is only of the fifth magnitude (at least, generally so; it is somewhat variable), so that with the naked eye it is not impressive even though it is actually redder than Betelgeux. Binoculars will show the colours of many stars, though the orange and red hues are the most pronounced.

Colour depends upon surface temperature. White heat is hotter than yellow, which in turn is hotter than red; thus the white Rigel in Orion is hotter than our yellow Sun, while the orange-red Betelgeux and Aldebaran are decidedly cooler. The surface temperatures range from below 3000 degrees Celsius for the coolest red stars up to well over 50 000 degrees for the very energetic stars which are white or bluish.

This is not a book about theoretical astronomy, but I must say something about the way in which a star radiates, because it is very much part of the over-all story. First, the source of stellar energy is nuclear. Normal stars contain a great deal of hydrogen, the lightest and most abundant substance in the whole universe (indeed, hydrogen atoms outnumber the atoms of all other elements put together). In a star's central regions, the nuclei of hydrogen atoms are bonding together to form nuclei of the next lightest element, helium. It takes four hydrogen nuclei to make one nucleus of helium, and each time this happens there is a slight release of energy and a slight loss of mass. It is this energy which keeps the star shining. The Sun is certainly 5000 million years old, and by stellar standards it is no more than middle-aged; not for another 5000 million years at least will anything drastic happen to it, so that we on Earth have a long reprieve (unless. of course, we decide to indulge in another major war, which would not destroy the world but would certainly wipe out the human race).

According to present-day ideas, a star begins its career by condensing out of a cloud of dust and gas known as a nebula. It shrinks, under the influence of gravity, and becomes hot inside. With a star of very low initial mass, nothing more happens; the star glows feebly, and then simply cools off until it has become dead. With a solar-type star, however, the temperature rises to ten million degrees or so, and nuclear reactions are triggered off, so that the star settles down to a long period of stable existence. Eventually, of course, the supply of available hydrogen 'fuel' must be used up, and the star will have to rearrange itself.

The sequence of events depends mainly upon the original mass. If the mass is about the same as that of the Sun, the core will shrink when the hydrogen runs low, while the outer layers will expand and cool. Different kinds of reactions start, and the star swells out to become a Red Giant, as Arcturus is today. There is a period of instability, but finally the outer layers are puffed away, and the star becomes a planetary nebula – a bad term, because a planetary nebula is not truly a nebula and has nothing whatsoever to do with a planet. Gradually the expelled layers move off and fade away, while all that is left of the star is the small, dense core, which is at a high temperature and is amazingly 'heavy'. Bankrupt stars of this kind are known as White Dwarfs. The tremendous density is due to the fact that all the parts of the atoms are crushed together with almost no waste of space. You could pack a ton of White Dwarf material into an eggcup.

After an immensely long period of feeble luminosity, the star loses the last of its light and heat, and becomes a cold, dead Black Dwarf. There is a certain doubt as to whether the universe is old enough for any Black Dwarfs to have been produced as yet, but eventually it must happen, and this will be the final fate of the Sun – though we will not be there to see; the Earth can hardly expect to survive the Red Giant stage, when the Sun will radiate at least a hundred times as fiercely as it does at present.

With a star of greater mass, everything happens at an accelerated rate, and the star's active life is much shorter than with our sedate Sun. As before, the star condenses out of a nebula, heats up, and begins to shine by the hydrogen-into-helium process. When it has exhausted its reserves, it collapses suddenly. There is a catastrophic 'implosion', which may be called the opposite of an explosion, followed by a shock-wave which literally blows the star apart in what is called a supernova outburst. The end result is a cloud of expanding gas, in the midst of which there may be a very small, super-dense object known as a pulsar, made up of neutrons – that is to say, atomic particles with no electrical charge. Because the neutrons can be so closely packed, the density is far greater even than that of a White Dwarf. This time, our eggcup could contain thousands of millions of tons.

This seems bizarre enough, but what about a star which is more massive still? When the grand collapse starts, it is so violent and so rapid that nothing can halt it. The star goes on becoming smaller and smaller, denser and denser, until its gravitational pull has become so enormous that not even light can escape from it. Light, of course is the fastest thing in the universe, and so the collapsed star has cut itself off. It has surrounded itself with a region from which nothing – absolutely nothing – can escape. It has become a Black Hole.

Black Holes are fascinating, and have caused a great deal of excitement over the past 30 years. It is tempting to delve further into the subject; but I am afraid that it would take me far outside my main theme, and so, reluctantly, we must leave matters there.

No star will appear as anything but a dot of light. If you use binoculars or a telescope and see a star as a large shimmering globe, you may be sure that there is something wrong with the focus. Therefore, most of our knowledge of the make-up of the stars has been drawn from instruments based upon the principle of the spectroscope.

Just as a telescope collects light, so a spectroscope splits it up. Pass a ray of sunlight through a glass prism (as Isaac Newton did, as long ago as 1666), and you will find that it is spread out into a rainbow band of colours, with red at one end and violet at the other. Light is a wave motion, and the colour of the light depends upon the wavelength – that is to say, the distance between one wave-crest and the next. Red light has the longest wavelength and violet the shortest, with orange, yellow, green and blue in between. If the radiation has a wavelength longer than that of red light, it cannot be seen, but it can be detected as infra-red and, with increased wavelength, radio radiation. To the short-wave end of violet we have ultra-violet, X-rays, and the extremely short, highly-penetrating gamma-rays. Visible light occupies only a very small part of the total range of wavelengths, or 'electromagnetic spectrum'.

If you look at the spectrum of a luminous solid, liquid or high-pressure gas, you will see the familiar rainbow. Look at the spectrum of a low-pressure luminous gas, and there will be no rainbow; instead there will be isolated coloured lines. Thus sodium, one of the two elements making up common salt, produces two bright yellow lines (as well as a host of others). Each element or group of elements has its own particular trademark, which cannot be copied, so that these two yellow lines must be due to sodium and nothing else.

Now consider the spectrum of the Sun. The bright solar surface is made up of gas at reasonably high pressure, so that in a spectroscope it yields a rainbow. Above it is a layer of gas at lower pressure. Normally this would produce bright lines; but because these lines are silhouetted against the rainbow background, they appear dark instead of bright. Their positions and intensities are not affected, so that they can be tracked down. Thus in the yellow part of the Sun's rainbow band there are two dark lines which correspond exactly to the two bright yellow lines of sodium, and we can prove that there is sodium in the Sun. By now over 70 elements have been identified in the stars, and no doubt the remainder will be found eventually.

It was the German optician Joseph von Fraunhofer who first studied the Sun's spectrum in detail, in 1814, which is why the dark lines – properly termed absorption lines – are often called Fraunhofer lines. Later, the spectra of other stars were examined, and before the end of the last century astronomers at Harvard College Observatory in the United States had drawn up a system of classification which is still used. The spectral types were given letters of the alphabet, as follows:

O: very hot stars, greenish-white or bluish-white. O-stars are rare.
B: hot white stars, such as Rigel in Orion.
A: cooler white stars, such as Sirius.
F: slightly yellowish. The Pole Star is of type F, though I doubt whether anyone will detect yellowness in it (I certainly cannot).
G: yellow stars, such as the Sun.
K: orange stars, such as Arcturus in Boötes (the Herdsman).
M: orange-red stars, such as Betelgeux.

There are a few extra types (W, hotter than O; and R, N and S, cooler than M) but these are so uncommon that for the moment we need not worry about them. The vast majority of stars can be divided into types B to M. Each type can be again subdivided; for example, a star of type K5 is intermediate between K (or K0) and M.

(You may wonder why the alphabetical sequence is so chaotic. The reason is that after the first attempts had been made, some types, such as C, D and F, were found to be unnecessary. The usual mnemonic is **O** Be **A** Fine **G**irl **K**iss **M**e.)

In 1830 the French philosopher August Comte wrote that 'the chemistry of the stars' was something which mankind could never learn. The fact that we now know so much about it proves that he was utterly wrong. Moreover, studies of the spectra of the stars can tell us a great deal about their real luminosities, and hence about their distances; once you know how powerful an object really is, and compare it with its apparent brilliancy, you can work out how far away it must be, provided that due allowance is made for complicating factors such as the absorption of light by material spread thinly through space. Some stars are true cosmic searchlights. Eta Carinæ, in the far south of the sky, can match at least six million Suns, though it is so remote that you need binoculars to see it. Rigel in Orion has 60 000 times the luminosity of the Sun, and even at its distance of 900 light-years it still shines as the seventh apparently brightest of all the stars; its magnitude is 0.1.

Using binoculars, it is possible to see the hues of many stars. Yet it is true that unless a star has a marked colour, there is nothing to single it out. It is the stars showing special characteristics which are of particular interest to the observer who has equipped himself with binoculars or a telescope.

4

Double Stars and Variable Stars

Fortunately for us, the Sun is a single star. If it had a companion, the climate of any of its planets would be decidedly uncomfortable. Yet double, triple and even multiple stars are surprisingly common in space, and some of them are within the range of binoculars or even the naked eye.

Probably the most famous of all pairs is Mizar, the second star in the 'handle' of the Plough; its official designation is Zeta Ursæ Majoris. Close beside it is a much fainter star, Alcor, which is an easy object when seen against a dark sky, and can be made out with no trouble at all. In some cases the two components of a double are not genuinely associated, and merely happen to lie in almost the same line of sight as seen from Earth. Take, for instance, Theta Crucis in the Southern Cross. There are two stars of between the fourth and fifth magnitudes, one rather brighter than the other, making up a wide naked-eye pair. The brighter star is 55 light-years away from us; the fainter as much as 950 light-years! There are no 3-D effects in space.

Yet rather surprisingly, most of the double stars are physically-associated or binary systems. This was established as long ago as the year 1802 by William Herschel, one of the greatest observers who has ever lived. The discovery was more or less accidental, because Herschel was carrying out a completely different kind of investigation.

He was trying to measure star distances by the method which is known as parallax. The best way to explain this is to make a simple experiment. Hold up a finger at arms-length, close one eye, and line your finger up with a picture or some other convenient object some way away. Now, without moving your finger, use the other eye. Your finger will no longer be aligned with the picture - because you are looking at it from a slightly different direction; your eyes are not in the same place. Measure the angular shift (the parallax), and also the distance between the two observing points (your two eyes). You can now use straightforward mathematics to work out the distance between your finger and your face. Surveyors use exactly the same principle when measuring the distance of an inaccessible object such as a mountain-top, though of course the base-line between the two observing points must be much greater than that between your eyes.

Herschel reasoned that a relatively nearby star would show a parallax shift against the background of more remote stars if it were observed from opposite sides of the Earth's orbit. Between January and June, the Earth moves from one side of the Sun to the other. The distance involved is 300 000 000 kilometres, twice the Earth's distance from the Sun. Therefore, Herschel studied double stars, expecting that the closer member of the pair would show a regular parallax shift at against its companion.

In theory he was correct, and in 1838, well after Herschel's death, the German astronomer F. W. Bessel managed to measure a star distance in this way. 61 Cygni, a faint star in the Swan, turned out to be about 11 light-years from us.* But Herschel's equipment was not sensitive enough, and in this particular investigation he failed. What he did find, to his surprise, was that some of the doubles moved in a way which indicated that they were in orbit round the 'balancing point' or common centre of gravity of the system. Castor, in Gemini (the Twins) was one such binary.

Today, thousands of binaries are known. Some are so widely separated that their motion relative to each other is too small to be measured at all, and all we can really say is that they are travelling through space together, at the same rate and in the same direction. The Mizar–Alcor pair is of this type, though it must be added that a telescope shows Mizar itself to be made up of two rather unequal components. Other binaries have shorter periods; for example 80 years in the case of the brilliant southern Alpha Centauri, which is a mere 4.5 light-years away from us. Also there are many binaries in which the components are so close together that no telescope will separate them, and they betray their true nature only by means of the spectroscope. One dim system in the constellation of the Hunting Dogs has proved to have a revolution period of only 17 minutes. Needless to say, no telescope will show it as anything but a single speck of light.

The apparent separation between the components of a double star is given in seconds or minutes of arc. A full circle is divided into 360 degrees, each degree into 60 minutes and each minute into 60 seconds, so that one second of arc is a very small angle indeed. It is worth remembering that the distance between the two Pointers in the Great Bear, Dubhe and Merak, is about 5 degrees. The angular separation between Mizar and Alcor is 700 seconds (approximately 11½ minutes of arc). In the case of Epsilon Lyræ, near the brilliant blue Vega, the separation between the two almost equal components is 208 seconds. These pairs are separable with the naked eye, but closer binaries – or, of course, optical doubles – require binoculars or a telescope.

I have made some tests, and it may help to give the results, though observers with keener eyes than mine will be able to do better. Nu Draconis, in the head of the Dragon, consists of two fifth-magnitude stars separated by 62 seconds of arc. I have never been confident that I can split them with the naked eye, but 7 × 50 binoculars make it easy enough. The fact that the two members of the pair are equal, and not very bright, is a great help. I have also looked at Beta Cygni or Albireo, in the Swan, where there is a golden yellow primary of magnitude 3.2 together with a blue companion of magnitude 5.4. The separation is 55 seconds of arc, and in any small telescope the two make a lovely spectacle; I always regard Albireo as the most beautiful double in the

---

* 61 Cygni has no Greek letter – after all, there are only 24 letters in the Greek alphabet. Numbers were allotted by the first Astronomer Royal, John Flamsteed, and are still used. Thus Rigel, or Beta Orionis, can also be known as 19 Orionis.

entire sky. Under ideal conditions I can just see the components separately with 20 × 70 binoculars, but I am not confident that I can do so with any lower magnification. In cases where the primary is very much the brighter of the pair, the secondary is effectively 'drowned'. Therefore, in the description of the constellations here, I have limited myself to pairs where the components are not too unequal, and are at least 35 seconds of arc apart.

It used to be thought that a binary resulted from the fission or breaking up of a formerly single star which was spinning rapidly, and became unstable. This idea has long since fallen out of favour; it is much more likely that the two components of a pair were born at the same time and in the same region of space, from the same cloud of dust and gas.

In 1669 the Italian astronomer Montanari noticed something very peculiar about the star Algol in Perseus. Normally it is of the second magnitude, just about equal to the Pole Star, but every 2½ days it gives a long, slow wink, taking four hours to fade down below the third magnitude and remaining at minimum for a mere 20 minutes before starting to recover – after which nothing more happens for the next 2½ days. Montanari was no doubt puzzled, and it was not until 1782 that the cause of Algol's variations was explained by John Goodricke, a most unusual astronomer inasmuch as he was deaf and dumb and died at the early age of 21.

Goodricke realized that Algol is not genuinely variable. It is a binary, made up of two components which are unequal in luminosity. When the fainter member passes in front of the brighter, it naturally cuts off some of the light, and Algol 'winks'. It is often called an eclipsing variable, though it should more properly be described as an eclipsing binary. Other stars of the same kind were found, and by now many are known. Of course, everything depends upon the angle from which we view them. We happen to lie in the line of sight so far as Algol is concerned (or nearly so); if we were observing from a different vantage point there would be no eclipses, and Algol would shine steadily.

With Beta Lyræ, close to Vega, things are different. There are two minima, one deep and one shallow, which occur alternately. This is because the two components are so close together that they almost touch, and presumably gravitational strains mean that each is distorted into the shape of an egg. The whole system is believed to be surrounded by streamers and shells of gas, and if seen from close range would be truly fascinating, though from its distance of 300 light-years it appears single in our telescopes.

Other stars are genuinely variable; they swell and shrink, changing their output as they do so. They are of various types, but at the moment we need consider only those which have representatives within binocular range.

1. **Cepheids**. Named after Delta Cephei, the first-discovered member of the class (by Goodricke, in 1784). These have short periods, from a few days to a few weeks. They are F or G giants, so that they can be seen over great distances. Cepheids are absolutely regular, so that their magnitude at any moment can always be predicted, and they are very useful to astronomers, because their real luminosities are linked with their periods; the longer the period, the more powerful the star. Once a star's real brightness is known, its distance can be worked out, so that the Cepheids act as 'standard candles'. It

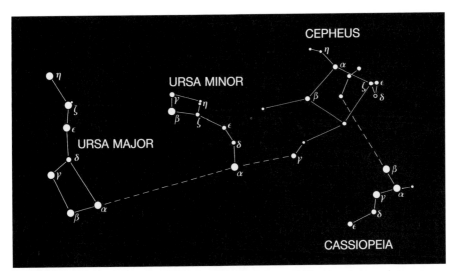

δ **Cephei**. *How to find δ Cephei using Ursa Major and Cassiopeia as pointers.*

was by studying short-period variables in the objects once known as spiral nebulæ that Edwin Hubble, in 1923, was able to prove that the spirals are external systems, far beyond the limits of our own Milky Way Galaxy.

There are several Cepheids well within binocular range, and some can be followed throughout their variations with the naked eye: Eta Aquilæ, Zeta Geminorum and Beta Doradûs, for example. In the far south we have Kappa Pavonis in the Peacock, which is what is termed a 'Type II Cepheid', rather less powerful than a classical Cepheid with the same period.

2. **Mira stars**. These are named after Mira Ceti in the Whale, which can occasionally become brighter than Polaris. The periods are long, from about 80 days to over 600 days, and the magnitude ranges are much greater than with the Cepheids. There is no law linking period with luminosity, and neither the periods nor the amplitudes are constant; thus Mira can sometimes pass through maximum without rising above the fourth magnitude. Virtually all Mira variables are Red Giants of types M, R, N or S.

Very few of these long-period stars are suitable for observation with binoculars. Even Mira is within range for only a month or two out of its total period of 531 days; at minimum it sinks to magnitude 10, so that it is lost even with a small telescope. The only other Mira stars which can reach the fourth magnitude are Chi Cygni, in the Swan, and R Hydræ, in the Watersnake.

However, there are several Mira variables which can be found with binoculars when near maximum, and can even show some colour; I have given notes about them in the pages which follow, but it is rather pointless to go into detail, because long-period variables are the province of the telescopic observer, and estimates made with binoculars are inevitably rough.

3. **Semi-regular variables**. Again most of these stars are red, with spectral types of M, R, N or S. Their amplitudes are less than for the Mira stars, and usually amount to little more than a magnitude. In some cases there are

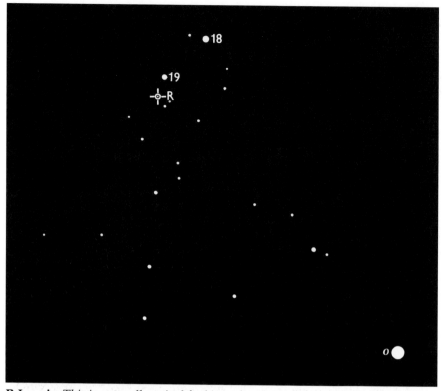

R Leonis. *This is not really suited for binocular observation, but it is one of the few Mira variables easily found with binoculars when near maximum. The period is 313 days, and the range from 5.4 to 10.5, so that for much of the time it is out of binocular range. The bright star to the lower right is Omicron Leonis (3.5). The only binocular comparison stars for R are 19 Leonis (6.4) and 18 Leonis (5.8).*

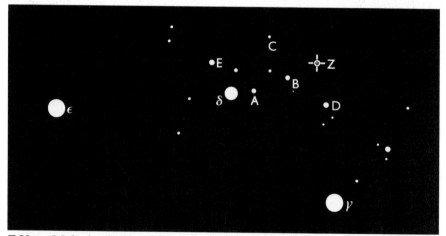

Z Ursæ Majoris. *A red semi-regular variable in the Great Bear, with a range from 6.8 to 9.1 and a rough period of 198 days; for part of the time it is an easy binocular object. The arbitrarily-lettered comparison stars given here are: A = 6.5, B = 7.2, C = 7.6, D = 5.9, E = 5.7.*

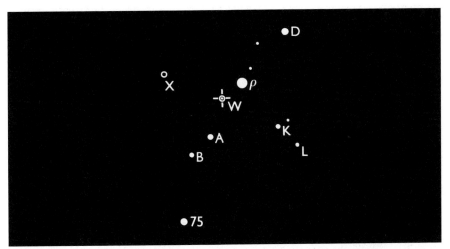

W Cygni. *A red semi-regular variable; range 5.0 to 7.6, period said to be about 130 days though I find little evidence of this. The variable is near ρ (4.0). Comparison stars are 75 Cygni (5.0), D (5.4), A (6.1), B (6.7), K (6.8) and L (7.5). Avoid the star marked X, which is itself variable. (Note that the letters A, B, D, K, L and X on this chart are unofficial and are used by variable star observers for convenience. For instance, the official X Cygni is near λ.)*

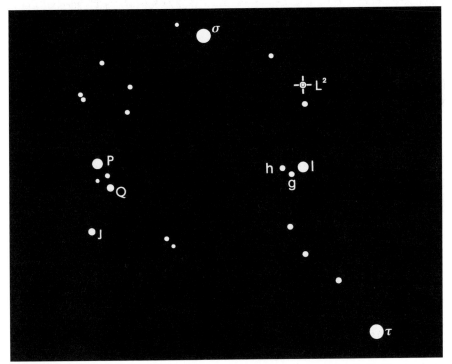

L² Puppis. *A red semi-regular variable, not far from σ (3.2) and τ (2.9). The range is from 3.4 to 6.2, and the period about 141 days. Comparison stars are I (4.5), P (4.1), Q (4.7) and J (4.6). Also uses the arbitrarily-lettered g (5.8) and h (6.2).*

definite indications of periodicity; thus Beta Pegasi, in the Flying Horse, has a period of about 36 days, though the range is less than half a magnitude. With other semi-regulars, the periods are so ill-defined as to be barely recognizable, and sometimes the fluctuations become random. The brightest of these variables is Betelgeux in Orion, which can become almost equal to Rigel, though normally it remains between magnitude 0.4 and 0.7. It has been said that there is a period of about 5½ years, but it is very vague indeed. A number of semi-regular variables can be followed with binoculars; R Lyræ and W Cygni, for example.

4. **Eruptive variables**. There are various types. With the RV Tauri stars there are alternate deep and shallow minima, with spells of complete irregularity. RV Tauri, the prototype star, is never bright enough to be seen with binoculars, and the only binocular representative is R Scuti, in the little constellation of the Shield.

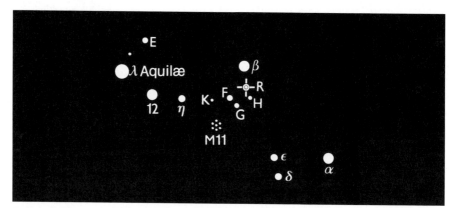

R Scuti. *This is an interesting star, and easy to follow with binoculars for almost all the time. The usual range is from 5 to just below 6. It is said to drop at times to below 8, but I have never yet lost it with 7 × 50 binoculars. It is easily found near the cluster M11; λ and 12 Aquilæ act as guides but both are too bright to be useful comparison stars, and so is α Scuti. The useful comparisons are all in Scutum: β (4.5), δ (4.9 very slightly variable), η (5.0), ε (5.1) and then the arbitrarily-lettered E (5.6), F (6.1), G (6.8), H (7.1) and K (7.7). R Scuti forms a quadrilateral with F, G and H, which is very easy to identify.*

5. **Irregular variables**. Again there are several different types. R Coronæ Borealis stars remain at maximum for most of the time, but suffer sudden, unpredictable drops to minimum, apparently because carbon particles – in other words, soot – accumulate in their atmospheres, and dim the light before being blown away. They are rare, and are luminous supergiants; R Coronæ itself is much the brightest of them, and is on the fringe of naked-eye visibility when at maximum. The red irregular variables may not be essentially different from the semi-regulars, but have never shown any real signs of periodicity, however rough. Mu Cephei the 'Garnet Star', is a famous example.

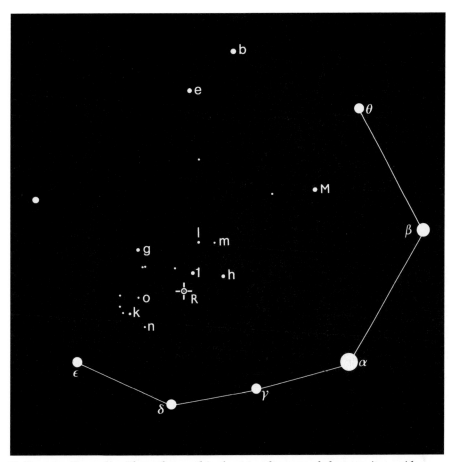

R Coronæ Borealis. *I have drawn this chart to a larger scale because it provides an excellent check of the limiting magnitude of binoculars. Generally R Coronæ is around magnitude 6. When it drops to minimum it may fall to 15, and will be out of binocular range for some time. The comparison stars given here (lettered arbitrarily) are: b = 5.6, M = 6.6, e = 6.7, 1 = 7.8, g = 7.6, h = 7.9, k = 8.3, l = 8.6, m = 8.9, n = 9.2, o = 9.5.*

6. **Unclassifiable stars**. These are decidedly rare. Gamma Cassiopeiæ in the W-shaped northern constellation of Cassiopeia is usually about magnitude 2.2, but in 1936 it flared up to 1.6 before fading back to below its normal brightness and then making a slow recovery. Rho Cassiopeiæ, in the same constellation, fluctuates around magnitude 5 but very occassionally drops to below 6. P Cygni, in the Swan, flared up from obscurity to magnitude 3 in 1600, and then declined; for well over a century now it has hovered around the fifth magnitude, easy to estimate with binoculars. The most erratic variable of all is the far southern Eta Carinæ, which for a while during the nineteenth century was the brightest of all stars apart from Sirius, but is now below naked-eye visibility, though binoculars show it well.

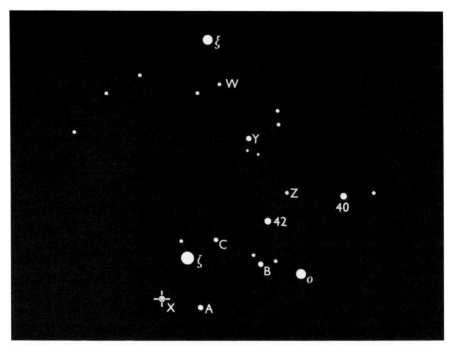

X Persei. *An interesting variable, and is a source of X-rays. It is suited to binocular observation. The range is about 6.0 to 6.6, and no period is known. Comparison stars are: 40 = 5.0, 42 = 5.1, Y = 5.7, A = 6.1, B = 6.2, C = 6.6. X is easily found, near ζ Persei. The letters for the comparison stars shown here are arbitrary but convenient.*

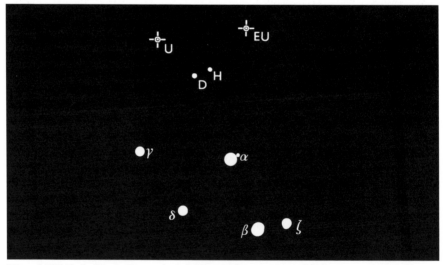

U and EU Delphini. *Two red variables. U has a range of from 5.6 to 7.5 and seems to be irregular. EU has a range of from 6 to 7 and is said to have a rough period of 60 days, though I have never found much evidence of it. Comparison stars are D (6.3) and H (6.8).*

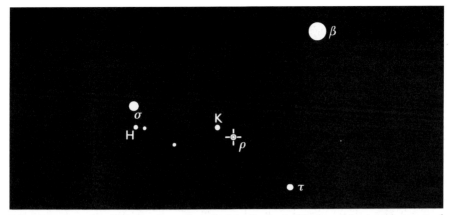

Rho Cassiopeiæ. *This has been described in the text. We do not know what sort of a variable it is; it does not seem to fit into any category. The extreme range seems to be from 5 to 6.4 or so, but generally it remains between 4.6 and 5.2; it has not been through a major minimum since 1949. Comparison stars are ρ (4.9), τ (5.1) and when a minimum occurs, K (6.0); τ has been suspected of slight variability. The variable is easily found, close to β.*

7. **Novæ**. These are stars which flare up suddenly from near-invisibility to prominence, remaining bright for a few days, weeks or months before fading again. They cannot be predicted, and amateurs have a fine record in nova hunting.

Apparently a nova is a close binary system, made up of a cool, normal star and a White Dwarf. The White Dwarf pulls material away from its companion, and builds up a shell around itself; when enough hydrogen has accumulated there is a violent though short-lived outburst.

When a nova is found, it is important to follow it as carefully as possible, and binoculars can be very helpful indeed, particularly as most bright novæ remain within binocular range for some time. HR Delphini, discovered by George Alcock in 1967, was not lost to binocular users until 1971. On the other hand Nova Cygni 1975, which rose to magnitude 1.8 in only a few hours, dropped below naked-eye visibility in less than a week, and by now I have lost sight of it even with the 39-cm reflector in my observatory.

Novæ generally appear in regions close to the Milky Way, and anyone who knows the sky really well has a chance of finding one, though beware of artificial satellites and even planets. (I remember having an urgent telephone call from a professional astronomer who believed that he had found a bright nova. It turned out to be Mars.) Only nine naked-eye novæ have appeared during the last twenty years, and there have not been many more within binocular range. A few stars have been known to undergo more than one major outburst; the 'Blaze Star', T Coronæ in the Northern Crown, is usually of about the tenth magnitude, but flared up to naked-eye visibility in 1866 and again in 1946. Stars of this type are known as recurrent novæ.

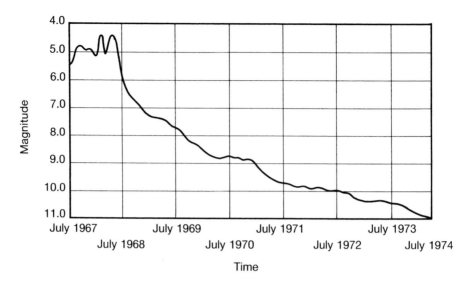

Light curve for a nova. *HR Delphini observed by Patrick Moore from discovery in 1967 to 1974.*

8. **Supernovæ**. These are colossal outbursts: at its peak a supernova may become at least 1000 million times as luminous as the Sun. In our Galaxy, only four have been seen during the last thousand years: in 1006, 1054, 1572 and 1604. The 1006 supernova, in Lupus (the Wolf) is believed to have become as brilliant as the quarter-Moon. The supernova of 1987, in the Large Cloud of Magellan, reached naked-eye visibility.

It is easy to draw up a light-curve for a variable star. What is done is to plot the observed magnitude against time. The method is to select at least two comparison stars, one brighter than the variable and the other fainter. For example: suppose that you have comparison stars A (magnitude 6 1) and B (6.7). If the variable is found to be 0.2 of a magnitude fainter than A but 0.4 brighter than B, then its magnitude must be 6.3. Sometimes the estimates are discordant, but a little practice will work wonders, and variable star work has become an important branch of modem amateur astronomy.

There are various complications. Many intrinsic variables are red, and it is not easy to compare a red star with a white one; it is wise to use more than two comparison stars if possible. It is in this kind of research that amateurs and professional astronomers work closely together; there are many variable stars in the sky, all with their own special points of interest. One never knows what will happen next!

5

Clusters, Nebulæ and Galaxies

Look into the evening sky at any time during northern winter or southern summer, and you cannot fail to notice the little group of stars of the Pleiades, otherwise known as the Seven Sisters. Normal-sighted people can see at least seven individual stars without optical aid, and binoculars raise the total to dozens. The chances of the grouping being accidental are effectively nil; we are dealing with a true star-cluster.

Open or loose clusters of this kind are common enough, and binoculars will show many of them. There are even several, in addition to the Pleiades, which can be seen with the naked eye; Præsepe in Cancer (the Crab) and the Jewel Box in the Southern Cross are other examples. It seems quite definite that the stars in a cluster have a common origin, and have remained as a group – though the gravitational pulls of non-cluster stars will be disruptive; eventually the cluster will lose its separate identity and will be dispersed, though the process is bound to take hundreds of millions of years at least. Some open clusters are relatively condensed; others are much more scattered. The Hyades, extending from the bright orange star Aldebaran in

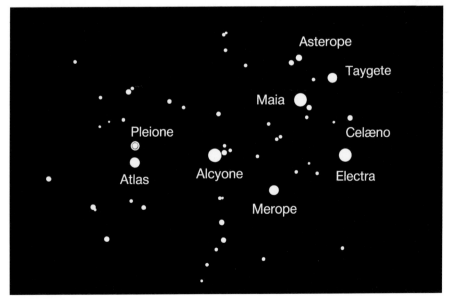

The Pleiades.

Taurus (the Bull) cover a much wider area than the Pleiades, and are also somewhat overpowered by Aldebaran, which does not actually belong to the cluster at all, but merely happens to lie about midway between the Hyades and ourselves.

With some of the open clusters, binoculars are adequate to show many individual stars, though in other cases the overall effect will be a gentle blur of light. Given sufficient magnification, of course, all open clusters can be resolved into stars. Very loose clusters such as the Hyades are best observed with low powers, when the entire cluster can be seen in the same field; fainter clusters require extra power. I know of many clusters which I can resolve with my × 20 binoculars, but not with × 12 or lower.

Open clusters are formless, though some of them are beautiful. I particularly like the Sword-Handle in Perseus, made up of two rich clusters in the same binocular field, and the Jewel Box in the Southern Cross, where most of the leading members are bluish but there is one red star which stands out. Globular clusters are quite different. They are vast, symmetrical systems, sometimes containing over a million stars each. They lie around the edges of the Galaxy, and all are thousands of light-years away, so that they appear faint. More than a hundred are known, though only three (Omega Centauri and 47 Tucanæ in the southern sky, and the Hercules globular in the northern) are clearly visible with the naked eye.

The globulars are not evenly distributed. There is a marked concentration in the region of Sagittarius (the Archer), and it was this which led to the discovery that we are having a lop-sided view of the Galaxy because we are not in the middle of it. The Sun, together with the Earth and the other planets, is around 30 000 light-years from the galactic centre, which lies beyond the glorious Sagittarius star-clouds.

With binoculars, few of the globulars can be resolved into stars except at their extreme edges. Omega Centauri, very much the brightest of them, is the only real exception (European and North American astronomers always regret that it is so far south). Toward the centre of a rich globular cluster the stars are separated not by light-years, but only by light-weeks or even light-days. If our Sun were a member of such a cluster the sky would be ablaze every night, with many stars brilliant enough to cast strong shadows. Moreover many of them would be red, because globular clusters are very old, and their leading stars have evolved into the Red Giant stage

Binoculars will show quite a number of globulars though in most cases only as tiny, blurred patches. They are worth seeking out, even though they are not spectacular unless observed with an adequate telescope.

There are several systems of nomenclature for clusters and nebulæ. In 1781 the French astronomer Charles Messier drew up a catalogue of them, containing over a hundred entries, not because he was interested in them (quite the contrary!) but because he was a comet-hunter, and he was persistently misled by fuzzy objects which looked like comets but which were not. Finally he decided to list them as 'objects to avoid'. His catalogue included not only clusters, both open and globular, but also nebulæ and external galaxies. The Messier or M numbers are still used. Thus the Pleiades cluster is M45, while the Hercules globular is M13. Later the Danish astronomer J. L. F. Dreyer compiled a more extensive catalogue, still known as the NGC or New General Catalogue even though it is now a century old. Many bright nebular objects are left out of Messier's list. In 1999 I compiled the Caldwell Catalogue, consisting of 109 bright objects omitted by Messier; rather to my surprise, the C numbers have come into general use. (I could not use M, of course, so I have used C; my surname is properly Caldwell-Moore.)

Gaseous or diffuse nebulæ, which we know to be stellar birthplaces, are in general rather disappointing binocular objects, though there are a few exceptions such as M42, the Sword of Orion (not to be confused with the Sword-Handle in Perseus). The usual trouble with nebulæ is that their surface brightness is low, and it is usually best to work with a comparitively low magnification. If I had to select one of my pairs of binoculars for observations of nebulæ I think I would choose the × 7 instrument, though I agree that the more powerful binoculars can bring out more details in the fainter nebulæ.

Planetary nebulæ, representing late stages in the evolution of normal stars, are decidedly elusive. Not many are accessible with binoculars, and to see the ring forms or the central stars means using a telescope. The few planetaries bright enough to be seen with binoculars show up in the guise of dim, ill-defined stars. The most celebrated of them is M57, the Ring Nebula in Lyra. I have often looked for it with binoculars, but I have never been able to glimpse it even with × 20, though my 76-mm refractor brings it out unmistakably, and with my 39-cm reflector I have no trouble in seeing the central star.

Finally, the Messier and Caldwell catalogues include many galaxies which are independent star-systems. Here again most of them are inconveniently faint, and in no cases can the forms be properly seen with binoculars. Even M31, the Great Spiral in Andromeda, looks like nothing more than an ill-defined misty patch, though its elongated shape is clear enough. Were it face-on to us, I suppose that binoculars would show the spiral; it is a great pity that it is placed at such an unfavourable angle.

However there are two exceptions: the Magellanic Clouds, so named because they were recorded by the Portuguese explorer Magellan in his voyage round the world – though they must have been noticed earlier, because they are very bright – and the Large Cloud is visible with the naked eye even in moonlight. They are not far from the south celestial pole, and they never rise over any part of Europe or the United States, but they are among the most familiar objects of the night sky to Australians, South Africans and New Zealanders. They are irregular systems. and may be regarded as companions of our Galaxy, since both lie within 200 000 light-years of us – as against over 2 000 000 light-years for the Andromeda Spiral.

The Clouds contain objects of all kinds: giant and dwarf stars, open and globular clusters, binaries, variables, novæ and gaseous nebulæ, one of which, the Tarantula Nebula, is a superb object in binoculars. If it were as close to us as the Orion Nebula, it would cast shadows. The Clouds are of immense importance to astronomers, and it is partly for this reason that many of the great new telescopes are being set up south of the equator, where the Clouds are accessible.

There was great excitement in February 1987, when a supernova flared up in the Large Cloud and became an easy naked-eye object. I saw it at its best, in April, when it was well above the third magnitude, and dominated the whole area; at that stage it was strongly orange-red. It faded slowly, and has now passed below binocular range.

Dedicated enthusiasts will be able to make themselves familiar with the brighter galaxies, and use binoculars to make regular surveys of them in the hope of catching sight of a supernova. This is certainly possible with telescopes, and one amateur in Australia, the Rev. Robert Evans, has more than twenty discoveries to his credit. With binoculars the chances are obviously much less, but one never knows – and it is true that in 1885 a supernova blazed out in the Andromeda Spiral and almost reached naked-eye visibility.

In the descriptions of the constellations which follow I have given hints on how to find the clusters, nebulæ and galaxies which are within binocular range. True, the low magnifications of ordinary binoculars mean that the scope is limited, but with open clusters, at least, binoculars are very suitable. The view of the Pleiades is much more breathtaking with my 7 × 50 binoculars than with the lowest power I can use on my large reflector, because the small field of view with the telescope means that only a small part of the cluster can be seen at any one time.

6

Stars of the Seasons

The 88 constellations officially recognized by the International Astronomical Union are very unequal in size and importance. The largest of them, Hydra, covers 1303 square degrees of the sky, the Southern Cross only 68 square degrees; Centaurus has 49 stars above the fifth magnitude, Mensa none at all. Before setting out on our binocular survey, it will be helpful to give a full list. In the course of a year the Sun passes through the 12 constellations of the Zodiac. In the table, 'northern' (N) means that the constellation is mainly in the northern hemisphere of the sky; 'far northern' (Far N) that it is always invisible, or inconveniently low, from latitudes such as that of South Africa or southern Australia; 'southern' (S) that the constellation is mainly south of the equator and 'far southern' (Far S) that it is inconveniently low, or invisible, from Britain or the northern United States. It is worth noting that Orion, one of the most splendid of all groups, is cut by the equator, though more of its really bright stars lie in the south than in the north.

I have also given the first-magnitude stars in each constellation. By convention, these include all stars down to magnitude 1.4. Thus Regulus (1.3) is said to be of the first magnitude, while Castor (1.6) is not.

Constellation		First magnitude star/s	Location
Andromeda	Andromeda	–	N
Antlia	The Airpump	–	Far S
Apus	The Bee	–	Far S
Aquarius	The Water-bearer	–	S Zodiacal
Aquila	The Eagle	Altair	N
Ara	The Altar	–	Far S
Aries	The Ram	–	N Zodiacal
Auriga	The Charioteer	Capella	N
Boötes	The Herdsman	Arcturus	N
Cælum	The Graving Tool	–	Far S
Camelopardalis	The Giraffe	–	Far N
Cancer	The Crab	–	N Zodiacal

Constellation		First magnitude star/s	Location
Canes Venatici	The Hunting Dogs	–	Far N
Canis Major	The Great Dog	Sirius	S
Canis Minor	The Little Dog	Procyon	N
Capricornus	The Sea-goat	–	S Zodiacal
Carina	The Keel	Canopus	Far S
Cassiopeia	Cassiopeia	–	Far N
Centaurus	The Centaur	Alpha Centauri, Agena	Far S
Cepheus	Cepheus	–	Far N
Cetus	The Whale	–	S
Chamæleon	The Chameleon	–	Far S
Circinus	The Compasses	–	Far S
Columba	The Dove	–	S
Coma Berenices	Berenice's Hair	–	N
Corona Australis	The Southern Crown	–	Far S
Corona Borealis	The Northern Crown	–	N
Corvus	The Crow	–	S
Crater	The Cup	–	S
Crux Australis	The Southern Cross	Acrux, Beta Crucis	Far S
Cygnus	The Swan	Deneb	Far N
Delphinus	The Dolphin	–	N
Dorado	The Swordfish	–	Far S
Draco	The Dragon	–	Far N
Equuleus	The Foal	–	N
Eridanus	The River	Achernar	Far S
Fornax	The Furnace	–	Far S
Gemini	The Twins	Pollux	N Zodiacal
Grus	The Crane	–	Far S
Hercules	Hercules	–	N
Horologium	The Clock	–	Far S
Hydra	The Watersnake	–	S
Hydrus	The Little Snake	–	Far S
Indus	The Indian	–	Far S
Lacerta	The Lizard	–	Far N
Leo	The Lion	Regulus	N Zodiacal

Constellation		First magnitude star/s	Location
Leo Minor	The Little Lion	–	Far N
Lepus	The Hare	–	S
Libra	The Balance	–	S Zodiacal
Lupus	The Wolf	–	Far S
Lynx	The Lynx	–	Far N
Lyra	The Lyre	Vega	N
Mensa	The Table	–	Far S
Microscopium	The Microscope	–	Far S
Monoceros	The Unicorn	–	Equatorial
Musca Australis	The Southern Fly	–	Far S
Norma	The Rule	–	Far S
Octans	The Octant	–	South polar
Ophiuchus	The Serpent-bearer	–	S
Orion	The Hunter	Rigel, Betelgeux	Equatorial
Pavo	The Peacock	–	Far S
Pegasus	The Flying Horse	–	N
Perseus	Perseus	–	N
Phœnix	The Phœnix	–	Far S
Pictor	The Painter	–	Far S
Pisces	The Fishes	–	N Zodiacal
Piscis Australis	The Southern Fish	Fomalhaut	S
Puppis	The Poop	–	Far S
Pyxis	The Mariner's Compass	–	Far S
Reticulum	The Net	–	Far S
Sagitta	The Arrow	–	N
Sagittarius	The Archer	–	Far S Zodiacal
Scorpius	The Scorpion	Antares	S Zodiacal
Sculptor	The Sculptor	–	Far S
Scutum	The Shield	–	S
Serpens	The Serpent	–	Equatorial
Sextans	The Sextant	–	Equatorial
Taurus	The Bull	Aldebaran	N Zodiacal
Telescopium	The Telescope	–	Far S
Triangulum	The Triangle	–	N

Constellation		First magnitude star/s	Location
Triangulum Australe	The Southern Triangle	–	Far S
Tucana	The Toucan	–	Far S
Ursa Major	The Great Bear	–	Far N
Ursa Minor	The Little Bear	–	North polar
Vela	The Sails	–	Far S
Virgo	The Virgin	Spica	Equatorial Zodiacal
Volans	The Flying Fish	–	Far S
Vulpecula	The Fox	–	N

(Some of these constellations have alternative names. Piscis Australis and Corona Australis may be Piscis Austrinus and Corona Austrinus respectively; Scorpius is often called Scorpio. Serpens is divided into two parts, Caput (the Head) in the northern hemisphere and Cauda (the Body) in the southern. Carina, Vela and Puppis are the dismembered parts of the old constellation Argo Navis, the Ship Argo.)

Obviously, one's view of the sky depends upon one's latitude. From Britain and the northern United States, for instance, Ursa Major is circumpolar – that is to say it never sets, and can always be seen whenever the sky is sufficiently dark and clear; from Hawaii it is not. Neither is it correct to believe, as many people do, that you have to go south of the Earth's equator to see the Southern Cross. It is easily visible from Hilo in Hawaii, where the latitude is 20 degrees north.

I have worked out a table which shows the visibility, or otherwise, of some of the brightest stars as seen from different places: C indicates that the star is circumpolar; V, that it is visible at times; and a dash, that it never rises.

Place	Polaris	Dubhe	Capella	Rigel	Canopus	Achernar	Acrux
Aberdeen	C	C	C	V	–	–	–
London	C	C	C	V	–	–	–
New York	C	C	V	V	–	–	–
Athens/ San Francisco	C	C	V	V	–	–	–
Hilo (Hawaii)	C	V	V	V	V	V	V
Darwin	–	V	V	V	V	V	V
Sydney/Cape Town/ Montevideo	–	–	V	V	V	V	C
Wellington	–	–	V	V	V	C	C
Falkland Islands	–	–	–	V	C	C	C

When considering the 'stars of the seasons' some allowances have to be made; obviously the view from, say, Britain and the northern United States will be different from that in Southern Europe or the southern United States; for example it is very difficult to see the Scorpion's 'sting' from London, but easy enough from Athens or San Francisco. However, I hope that these notes will be good enough to serve as a general guide. The seasonal charts are for late evenings, with the Sun far enough below the horizon for the sky to be dark. In general, on small-scale maps it is of course quite pointless to be too precise. The charts given here are I hope, detailed enough to show the over-all view. There is a difference in the positions amounting to 2 hours per month; thus the sky for midnight on 1 December is the same as at 4 hours GMT on 1 October, 2 hours GMT on 1 November, 22 hours GMT on 1 January, 20 hours GMT on 1 February, and so on. Naturally, the planets will show individual shifts during the same period. The brighter stars within each constellation are shown using larger symbols, but for accurate magnitudes reference must be made to the detailed charts in Chapter 7.

NORTHERN HEMISPHERE

Winter

During winter evenings Capella is almost at the zenith or overhead point, and stands out at once because of its brilliance; it is in fact the sixth brightest star in the sky, with a magnitude of only just below zero. Ursa Major, the Great Bear (nicknamed the Plough or the Big Dipper) is high in the north-east, with its curve pointing downwards. (It is tempting to call this curve 'the tail', though in fact the old mythological figures show it as the Bear's head.) Vega, in Lyra, is skirting the northern horizon, and any slight mist will hide it; it is worth noting that Vega, Polaris and Capella lie in a fairly straight line, so that when Capella is high up Vega is low down, and vice versa. The Square of Pegasus is dropping in the west, and Leo rising in the east. The southern aspect is dominated by Orion, which is quite unmistakable. The three stars of the Hunter's Belt point upward to the orange Aldebaran, in

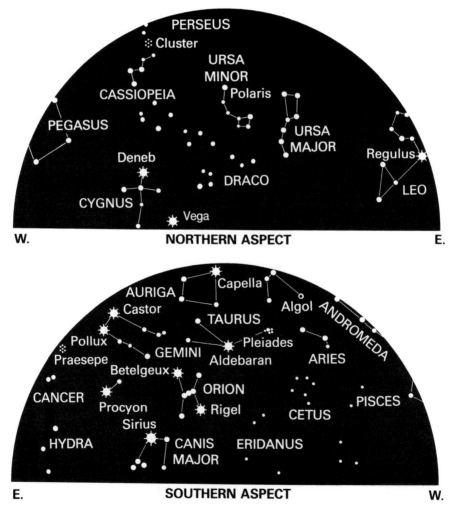

Taurus, and downward to Sirius, in Canis Major; other members of the Hunter's retinue include Procyon in Canis Minor and the Heavenly Twins, Castor and Pollux.

Spring

Orion is setting in the west; Sirius is out of view, though Procyon and the Twins remain, and Aldebaran is still above the horizon. The Great Bear is now practically overhead; Capella is fairly high in the west and Vega in the east, so that it is interesting to compare the two; in brightness they are almost equal, but Capella is a yellow star whereas Vega is vivid blue. The southern aspect is dominated by Leo, with the famous pattern making up what is termed the Sickle; to find it, use the Pointers, Dubhe and Merak in the Bear, in the direction away from Polaris. Follow round the Bear's curve and you will come to Arcturus in Boötes, which is a lovely light orange and is actually the brightest star in the northern hemisphere of the sky.

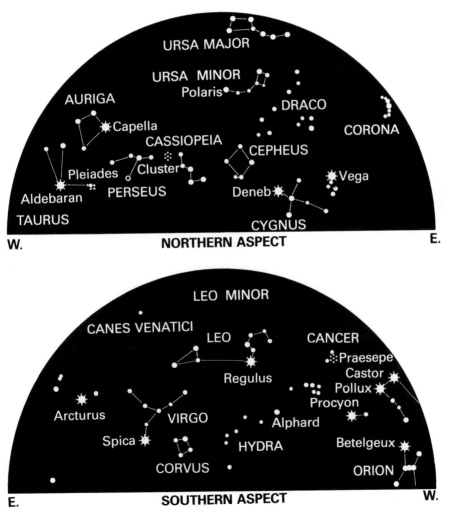

Summer

By summer evenings the Great Bear is descending in the north-west, while Leo has almost gone; the Square of Pegasus has come into view in the east. The sky is dominated by three bright stars: Vega in Lyra, Altair in Aquila and Deneb in Cygnus. Years ago, in a *Sky at Night* television programme, I referred to these three as making up the 'Summer Triangle', and nowadays everyone seems to use the term, though it is completely unofficial and in any case does not apply to the southern hemisphere, where June is midwinter. The charts given here are a little misleading, because Deneb has to be drawn on one section and Vega and Altair on the other, but the overall aspect is obvious enough. Arcturus is still high, and the lovely Zodiacal constellations of Scorpius and Sagittarius are to be seen low in the south. Scorpius is distinguished by the presence of the red supergiant Antares; in these regions, too, the Milky Way is particularly spectacular.

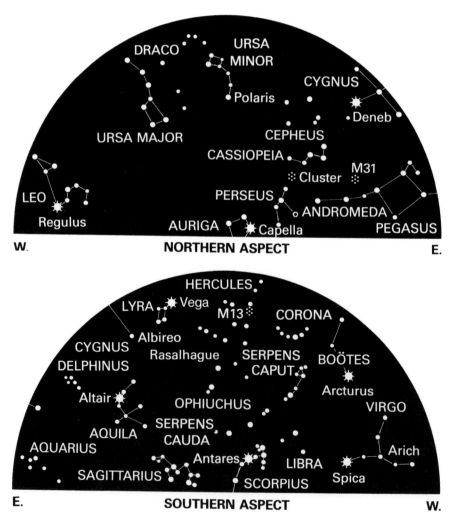

W.   **NORTHERN ASPECT**   E.

E.   **SOUTHERN ASPECT**   W.

Autumn

Autumn evenings are less brilliant. The Great Bear is at its lowest in the north, while the W of Cassiopeia, to the opposite side of the Pole, is near the zenith. Orion has yet to make his entry; we have lost Scorpius and Sagittarius, while Capella is rising in the north-east and Vega dropping in the north-west, though the other two members of the 'Summer Triangle', Deneb and Altair, are still high. The southern aspect is dominated by Pegasus, which is large but not so bright as might be thought from its aspect on the maps. Two of the stars in the Square point downward to Fomalhaut in Piscis Australis, which is the southernmost of the first magnitude stars to be visible from Britain; from South England and New York latitudes it can be seen easily enough, but from North Scotland it barely rises.

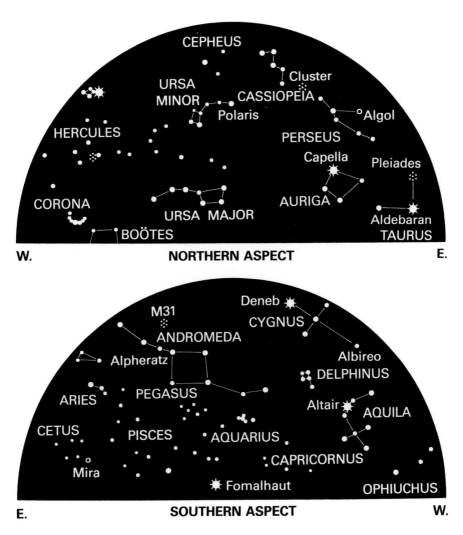

W. **NORTHERN ASPECT** E.

E. **SOUTHERN ASPECT** W.

SOUTHERN HEMISPHERE

Summer

For the southern maps, of course, everything is reversed with respect to the northern hemisphere, and in Orion we find that Rigel is to the top left of the main pattern and Betelgeux to the lower right. Orion is high in the north; the Belt points downward to Aldebaran and upward to Sirius. A line extended from Alnitak, the uppermost star of the Belt, through Saiph (Kappa Orionis) will show the way to Canopus in Carina, which is much the brightest of all the stars apart from Sirius. Capella is low in the north, with Leo grazing the horizon.

In the south you will find Crux Australis, the Southern Cross, which is not genuinely X-shaped; it is more like a kite, but it contains three stars of above the second magnitude. Below it are the two brilliant Pointers, Alpha and Beta Centauri; Alpha, the nearest of all the bright stars beyond the Sun, is well above zero magnitude so that only Sirius and Canopus outshine it. Centaurus almost surrounds the Cross, and until the seventeenth century the Cross was not even classed as a separate constellation.

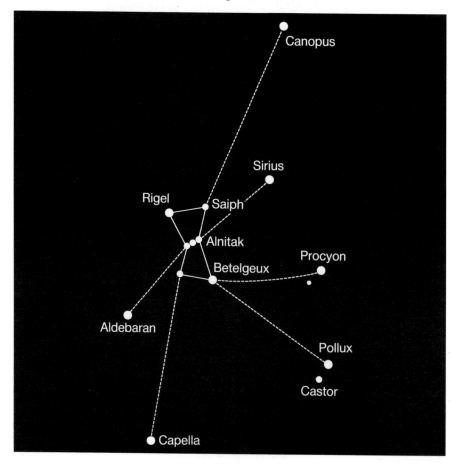

Locating the south celestial pole is not really difficult (I have given detailed directions on page 113), though there is no bright star close to it. The key is given by Achernar, now high in the south-west, which is easy to identify because of its brilliance and its relative isolation. Take a line from the longer axis of the Southern Cross and extend it as far as Achernar; the south pole lies about midway along it. The nearest naked-eye star to the pole is Sigma Octantis, but as it is below the fifth magnitude it will be hidden by mist or moonlight.

Look also for the two luminous patches of the Magellanic Clouds. The Large Cloud is so bright that it can be seen even when the Moon is obtrusive, and against a dark sky the Small Cloud too is conspicuous. The Clouds look superficially like detached parts of the Milky Way, but they are in fact independent galaxies.

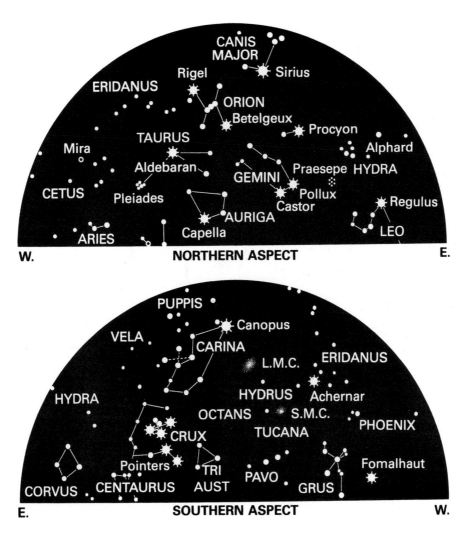

W.   **NORTHERN ASPECT**   E.

E.   **SOUTHERN ASPECT**   W.

45

Autumn

By autumn evenings Orion has set soon after darkness, but Sirius and Canopus are still high. This is the best time of the year for seeking out the Great Bear, though from most of Australia and South Africa it is always very low, and from Sydney or the Cape only part of it rises. (New Zealanders will lose it. Alkaid, the end star of the curve, just rises from Auckland, but it is very difficult to see.) Leo is high in the north, and Arcturus is well above the horizon. The Cross and the Centaur are now high in the south, so that Achernar is low down; Scorpius has come into view. The region near the zenith looks rather dull, because much of it is occupied by the vast, dim Hydra.

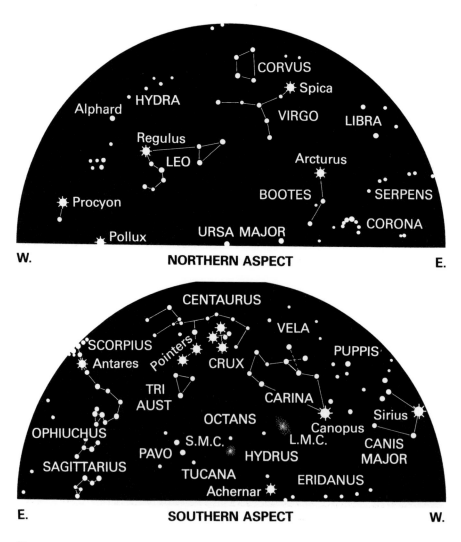

W. **NORTHERN ASPECT** E.

E. **SOUTHERN ASPECT** W.

Winter

Scorpius is now nearly overhead, and is one of the most magnificent of all constellations, particularly in view of the richness of the Milky Way here and in Sagittarius. The Cross is descending in the south-west, followed by the Pointers; Canopus is so low that it will probably not be seen, and from parts of Australia and South Africa it actually sets. Vega, Altair and Deneb are in the northern part of the sky, making up what should now be called the 'Winter Triangle'. Look also for Arcturus and Fomalhaut.

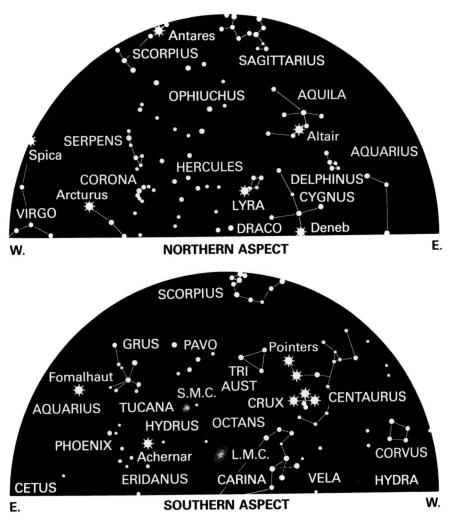

Spring

Sirius and Canopus have returned to the evening sky, though Sirius is still low in the south-east. Scorpius is setting, and the Cross is more or less out of view, though as seen from anywhere south of Sydney or the Cape it never drops below the horizon. The Square of Pegasus is high in the north, with two of its stars pointing upward to Fomalhaut close to the zenith. The Magellanic Clouds are prominent, and this is the best time to unravel the somewhat confusing region of the Southern Birds – Grus, Pavo, Tucana and Phœnix. The best method is to begin at the fainter of the two Pointers (Agena, or Beta Centauri) and extend a line from it through Alpha Trianguli Australe; prolonged for far enough this will lead to Alpha Pavonis in the Peacock, after which the other Birds can be sorted out.

I appreciate that these directions are somewhat rough and ready, but I hope that they will be useful. As soon as you have learned how to find the various groups, you can bring your binoculars into action, with results which will certainly not disappoint you.

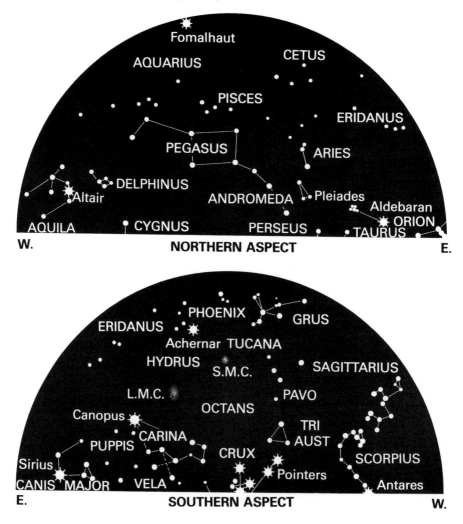

48

7

The Constellations

While writing this book I carried out a personal survey of the whole sky, using the binoculars which I have listed on page 11. I discovered that some of my results did not agree with those of other observers, and of course everything depends upon the quality of one's eyesight. It is also obvious that, say, 12 × 80 binoculars will have a much greater light-grasp than the 12 × 40 pair which I used, though there will be more of a problem in hand-holding.

It is a good idea to do some preliminary tests, particularly with regard to the field of view. My 12 × 40 binoculars will just take in the two Pointers to Polaris, Merak and Dubhe, which are 5 degrees of arc apart; the 7 × 50 pair will contain them very easily, while, naturally, 20 × 70 will not. Some useful standards are:

> Merak to Dubhe: 5 degrees.
> Alpha Centauri to Agena; 4½ degrees.
> Castor to Pollux: also 4½ degrees.
> Altair to Beta Aquilæ: 2½ degrees.
> Beta Arietis to Gamma Arietis: 1½ degrees.
> Average diameter of the full Moon: half a degree.
> Mizar to Alcor: 11.8 minutes of arc.
> Alpha^1 to Alpha^2 Capricorni: 6 minutes.
> Epsilon^1 to Epsilon^2 Lyræ: 3½ minutes.
> The components of Nu Draconis: 62 seconds of arc.
> The components of Beta Cygni: 35 seconds.

In charting the constellations I have added useful 'linking lines', though these are quite arbitrary. Very small constellations have not been given separate maps, and in each case I have added some of the stars in adjacent groups to help in identification. Magnitudes have been given to only one-tenth, because it is almost impossible for the naked-eye or binocular observer to distinguish differences of less than this though I agree that some people claim to be able to do so; I certainly cannot.

I do not suggest for a moment that I have described every interesting object within binocular range, but I think that there is enough variety here to satisfy most people. So let us now consider the constellations one by one, and see what we can find.

WHOLE SKY REFERENCE MAP

Northern Hemisphere

A NOTE ON THE MAPS

One trouble about writing a book intended for use in both hemispheres of the world is that what is the right way up to a Briton or a North American will be upside-down to an Australian or a South African, and vice versa. Unfortunately it is impossible to please everybody. What I have done is to give the maps an orientation which I have personally found to be convenient – I must apologize for a northern-hemisphere bias. But I hope that the different views will present no real problem.

WHOLE SKY REFERENCE MAP

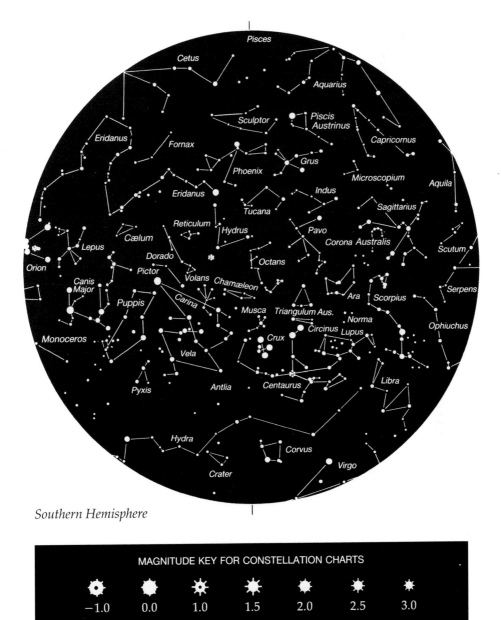

Southern Hemisphere

MAGNITUDE KEY FOR CONSTELLATION CHARTS

| −1.0 | 0.0 | 1.0 | 1.5 | 2.0 | 2.5 | 3.0 |

| 3.5 | 4.0 | 4.5 | 5.0 | 5.5 | 6 and under |

Variable stars

Bright and faint
nebulæ and clusters

ANDROMEDA: *Andromeda*

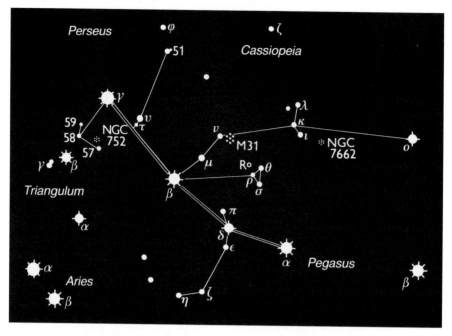

This is a large and important northern constellation, leading out from the Square of Pegasus in the direction of Capella. The chief stars are Alpha or Alpheratz (2.1), Beta (also 2.1), Gamma (also 2.1), Delta (3.3) and 51(3.6). Alpheratz is one of the four stars of the Square of Pegasus, and was formerly, and more logically, listed as Delta Pegasi. Beta is a fine orange star of spectral type M, beautiful in binoculars; it has been suspected of slight variability. Gamma, of type K, is light orange when seen in binoculars. and is a lovely binary, though the companion is too close-in to be seen with binoculars. Delta, also of type K is another orange star, and so is Xi (4.1).

The Mira variable R Andromedæ is in the × 20 field with the little triangle made up of Theta (4.6), Sigma (4.5) and Rho (5.2) It has a period of 409 days, and a very large range, from slightly above magnitude 6 at maximum down to almost 15 at minimum. For most of the time it is out of binocular range, but when at its brightest it is easy enough to find, and binoculars of × 8 or higher power will show colour in it; it is very red, and of spectral type S.

Of course the most famous object in Andromeda is the Great Spiral, M31, which has been known from very early times. It is clearly visible with the naked eye, close to Nu (4 5). Its distance is now known to be 2.2 million light-years, making it the closest of the really large galaxies. Unfortunately it must be said that in binoculars (or, for that matter, in most telescopes) it is disappointing, because it lies at a narrow angle to us and the full beauty of the spiral is lost. Even with × 20 it shows up as nothing more than an elliptical blur, more or less devoid of detail, though it is possible to see the smaller companion galaxy M32 close beside it. The other companion. NGC 205, is beyond binocular range; at least, I have never been able to see it without a telescope.

There is also the open cluster NGC 752 (C28), not far from Gamma, between it and Beta Trianguli. Locate the stars 59, 58 and 56; the cluster lies near 56, which is made up of a pair. It is not readily identifiable with × 12 or lower power, but is easy with × 20 though it is scattered and not at all conspicuous.

I have also looked for planetary nebula NGC 7662, which is in the field with the triangle made up of Lambda (3.8), Kappa (4.1) (C22) and Iota (4.3). I think I can just glimpse it as a dim fuzz, though large telescopes are needed to bring out its form and its very hot central star, which has a surface temperature of around 75 000 degrees

M31, the Great Spiral galaxy in Andromeda
(© Hale Observatories).

ANTLIA: *the Airpump*

One of the most obscure of all constellations; its brightest star, Alpha, is only of magnitude 4.2. It adjoins Vela and Pyxis, and is always very low as seen from Britain; part of it does not rise at all. It contains nothing of real interest, and I have included it in the chart with Vela.

APUS: *the Bee*

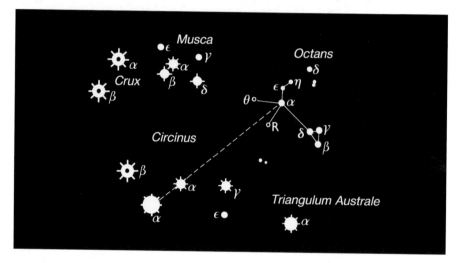

A far southern constellation, originally Avis Indica (the Bird of Paradise). To locate it, take a line from Alpha Centauri through Alpha Circini (which is in the same × 7 field) and continue until you come to Alpha Apodis (3.8). Once Alpha has been located, it is easy to find the other chief stars of the constellation, Gamma (3.9), Beta (4.2) and Delta (4.7), which make up a small triangle. With × 7, this triangle is in the same field with Alpha. Also in the field with Alpha are two much fainter stars, Eta (4.7) and Epsilon (5.2) which show the way to Delta Octantis, the 'key' to the south celestial pole.

Alpha itself is slightly yellowish, and is close to Theta, which is decidedly red; Theta has an M-type spectrum, and when at its brightest the colour is detectable even with × 7, though × 12 shows it better. It is variable, but never quite reaches naked-eye visibility; at minimum it is difficult with × 7. Delta is also red, and has a companion at a distance of 103 seconds of arc. Another red variable is R, in the same × 7 field with Alpha and Theta; the range is uncertain, but probably from around magnitude 5 to 6. The spectrum is of type M.

AQUARIUS: *the Water-bearer*

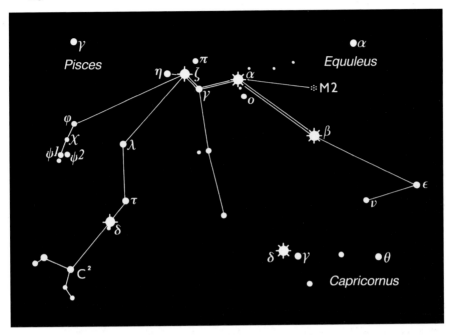

Aquarius is one of the larger but fainter constellations of the Zodiac. It can be identified by using Alpheratz and Alpha Pegasi in the Square of Pegasus as guides, and the line of stars of which Alpha Aquarii is the brightest member is easy to find. The leading stars are Beta (2.9), Alpha (3.0), Delta (3.3) and Zeta (3.6). Both Alpha and Beta are very distant, and are well over 5000 times as luminous as the Sun.

There is a distinctive small group of stars between Fomalhaut, in Piscis Australis, and Alpha Pegasi. The three stars lettered 'Psi Aquarii' are close together, with Chi and Phi nearby; several of them are orange. They have even been mistaken for a very loose cluster, though in fact they are not really associated with each other.

The main object of binocular interest in Aquarius is the globular cluster M2. I have never been able to see it with the naked eye (as some observers claim to have done), but it is not hard to find with × 7, and it is easy with × 8.5. The way to locate it is to begin at Beta and then sweep past a line of three faint stars until you come to the cluster, which, incidentally, forms a right angle with Alpha and Beta. It was discovered as long ago as 1746, and is some 50 000 light-years away, with a real diameter of about 150 light-years. I can resolve it with my 28-cm reflecting telescope, but even in × 20 binoculars it shows up only as a blur. Still, it is well worth finding, and it is one of the brightest globulars visible from northern latitudes.

AQUILA: *the Eagle*

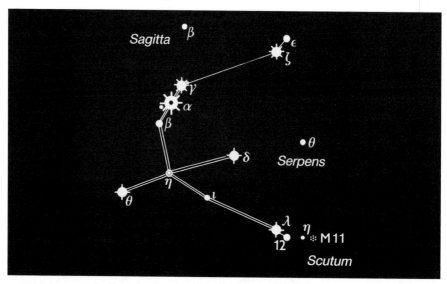

This is a large and splendid constellation which gives a vague impression of a bird in flight. The leading stars are Alpha or Altair (0.8), Gamma (2.7) and Zeta (3.0).

Altair, at a distance of 17 light-years, is one of the closest of the bright stars. It has ten times the luminosity of the Sun, and is pure white, with an A-type spectrum. It is one of the so-called Summer Triangle. Altair is flanked to either side by a fainter star, Gamma or Tarazed and Beta (3.7); Gamma is a K-type star, very clearly orange when seen in binoculars. The line of three makes Altair particularly easy to recognize. Antares in the Scorpion is also the centre of a line of three, but the colour-difference alone means that there can be no confusion; Antares is fiery red.

South of Altair there are three stars lined up: Theta (3.2), Eta (variable) and Delta (3.4). Eta is a Cepheid. It was identified as such only a short while after Delta Cephei itself, and if it had been found a few months earlier the short-period stars would probably have been known as Aquilids rather than Cepheids. Eta Aquilæ has a range of from 3.4 to 4.7, and a period of 7.2 days; Beta, Delta, Theta and Iota (4.4) are useful comparisons. Eta is 440 light-years away, and can attain a luminosity well over 5000 times that of the Sun.

Aquila ends to the south in a pair of stars, Lambda (3.4) and 12 (4.0). These two are the best guides to the little constellation of Scutum, with its famous open cluster M11; indeed, Scutum used to be included in Aquila, and there does not seem much justification for giving it a separate identity. The Milky Way runs right through Aquila, and is very rich, so that the whole region will repay sweeping with binoculars of any magnification.

Finally, several novæ have appeared in Aquila during recent years, so that it is always worth making a check – though do not be deceived by a slow-moving artificial satellite!

ARA: *the Altar*

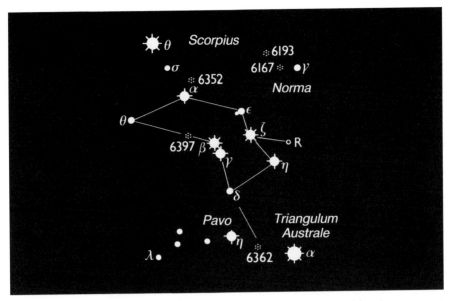

A far-southern constellation, lying between Theta Scorpii on one side and Alpha Trianguli Australe on the other. The chief stars are Beta (2.8), Alpha (2.9), Zeta (3.1) and Gamma (3.3). Beta and Zeta are orange; so is Eta (3.8). All three are of type K.

Ara has a fairly distinctive shape. It contains several clusters within binocular range; NGC 6397 (C86), 6362 and 6352 (C81) are globular, while NGC 6167 and 6193 (C82) are loose. Of these, the most notable is NGC 6397 (C86). It is quite easy to find, close to the Beta–Gamma pair; it is not particularly rich or condensed, but it may be only about 8200 light-years away, in which case it is the closest of all the globulars. I find NGC 6352 (C81) difficult with binoculars, even × 20.

R Aræ, in the same × 7 field with Zeta and Eta, is an Algol-type eclipsing binary with a period of 4.4 days. As its range is from 5.9 to 6.9, it is always easy to see with binoculars of any magnification.

ARIES: *the Ram*

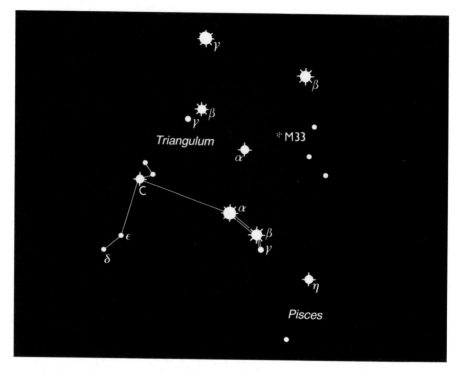

Aries is always classed as the first constellation of the Zodiac, though by now the First Point of Aries (the position where the ecliptic cuts the equator) has shifted into the adjacent constellation of Pisces. The three main stars in Aries are Alpha (2.0), Beta (2.6), c (3.6) and Gamma (3.9); Alpha, Beta and Gamma make up a conspicuous little group. Alpha, or Hamal, forms a large triangle with Beta and Gamma Andromedæ. It has a K-type spectrum, but I always find the colour less pronounced than with most bright K-stars, though binoculars show it to be decidedly 'off-white'. Gamma is a lovely telescopic double, with equal components, but as the angular separation is only just over 8 seconds of arc it is not resolvable with binoculars.

AURIGA: *the Charioteer*

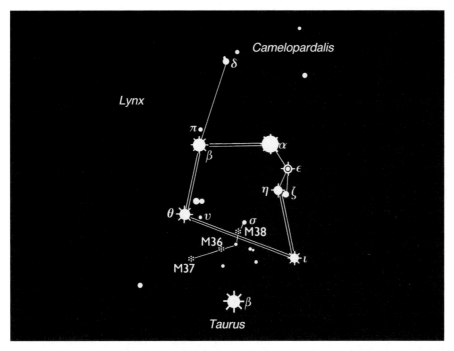

Auriga is one of the most striking of all the constellations of the northern hemisphere. Its leading stars are Alpha or Capella (0.1), Beta (1.9), Theta (2.6), Iota (2.7), Epsilon (3.0 at maximum), Eta (3.2) and Delta and Zeta (each 3.7, though Zeta is variable over a small range). Auriga has lost its former Gamma, or Al Nath, which has been transferred to the Bull as Beta Tauri. This seems rather illogical, because it seems to belong more naturally to the rather irregular box-shape of Auriga.

Capella is the sixth brightest star in the sky. It is 42 light-years away, and over 150 times as luminous as the Sun. Its declination is 46 degrees north, so that it is circumpolar from Britain; it can be seen from most inhabited countries, though it is always very low from New Zealand and does not rise at all from the latitude of Invercargill. It is yellowish, with a G-type spectrum, and is actually a close binary, though the components are so close together that no ordinary telescope will separate them.

Iota and Delta have K-type spectra, and are clearly orange when seen through binoculars. There are also two fainter M-type stars, Pi (4.3) in the × 7 field with Beta, and Upsilon (4.7) in the same field as Theta.

Close beside Capella there is a triangle of stars; Epsilon, Eta and Zeta, known collectively as the Hædi or Kids. They are in the same binocular field, even with × 20. Eta is a normal white B-type star, about 400 times as luminous as the Sun and 200 light-years away. The other two Kids are remarkable eclipsing binaries. They are not associated with each other; Epsilon is over 4500 light-years away, Zeta only a little over 500 light-years, so that their juxtaposition in the sky is pure coincidence.

AURIGA

Epsilon consists of an extremely luminous supergiant, perhaps over 100 000 times as powerful as the Sun, together with a mysterious companion which has never been seen, and is known only because it periodically passes in front of the supergiant and dims it by about a magnitude. The period is 27 years, and the eclipse lasts for a long time; the last began on 22 July 1982 and did not end until 25 June 1984, though it was total only for a year (January 1983 to January 1984). The nature of the secondary is still uncertain. It was once believed to be a very young star, still condensing; then it was suggested that we might be dealing with a Black Hole; now it seems more likely that the eclipsing component is a hot bluish star surrounded by a cloud of opaque material. Of course the fluctuations of Epsilon Aurigæ are noticeable with binoculars or even with the naked eye, at the time of eclipse; Eta makes a suitable comparison star. But it is also worth while to check up at other times, because there is still a great deal about Epsilon Aurigæ which we do not know, and it may not shine quite steadily at any time.

Zeta Aurigæ has a period of 972 days. Here we have a K-type supergiant, whose colour is very obvious in binoculars. The companion is a B-type star, 400 times as luminous as the Sun. Zeta Aurigæ is of great importance to astronomers, because before totality (that is to say, the moment when the blue star passes completely behind the red companion) begins the light of the blue star comes to us through progressively deeper layers of the supergiant's atmosphere and provides information about the composition and structure of the supergiant itself; but the magnitude range is only from 3.7 to 4.2.

The Milky Way flows through Auriga, and there are also some fine open clusters here. With × 7, there are three in the same field: M36, 37 and 38. The brightest of them, M37, is in the same × 7 field with Theta, but barely in the same field with × 12. With × 20 there is a hint of resolution into stars. M36 is usually given as slightly brighter than M37; to me it looks fainter, though more condensed, and again there is a hint of resolution with × 20. M38 is larger and looser, but less bright.

BOÖTES: *the Herdsman*

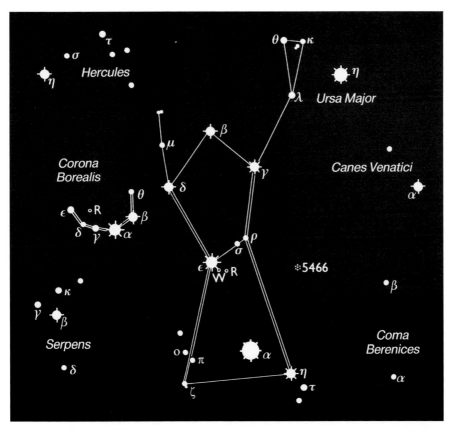

A very prominent constellation. It contains Arcturus, which is surpassed only by Sirius, Canopus and Alpha Centauri. Since the declination of Arcturus is 19 degrees north, it is sufficiently close to the equator to be seen from every inhabited continent. The magnitude is −0.04. The other leading stars of Boötes are Epsilon (2.4), Eta (2.7), Gamma (3.0), Delta and Beta (each 3.5) and Rho (3.6). The overall impression of the whole area is that of a distorted Y, with Alpha Coronæ Borealis included in the pattern. Arcturus can be found by following round the curve of Ursa Major, but it is in any case quite unmistakable; it has a K-type spectrum, and is a lovely light orange, so that in binoculars it is beautiful. It is one of the closer stars, at a distance of only 36 light-years, and is 115 times as luminous as the Sun.

Epsilon is also of type K, and the colour is obvious in binoculars; the same is true of Rho, which to me seems rather more highly-coloured than Epsilon. Using × 7, Rho is in the same field as Sigma (4.5) which is of type F, but in which I can see no colour at all. Zeta (3.8), Pi (4.9) and Omicron (4.6) make up a triangle in the same × 7 field; Omicron is decidedly orange. Tau (4.5) is known to be accompanied by a massive planet.

One star which shows obvious colour is W Boötis, which is of type M and is slightly variable with a range of from 5.0 to 6.4. It is in the same binocular

BOÖTES

field with Epsilon, and as it is the only other fairly bright star in the field it is impossible to miss. Binoculars bring out the orange-red hue clearly, though I never find it very pronounced.

One investigation is worth carrying out, though with scant hope of success. In April 1860 J. Baxendell, a leading variable star observer of the time, recorded an unfamiliar star of about magnitude 9.7 in the same field as Arcturus. It gradually faded away, and before the end of the month it had disappeared. It has never been seen again, though it has been given an official variable star designation: T Boötis. It may have been a recurrent nova, and although it is probably unlikely that it will ever come within binocular range, if it reappears at all, there is no harm in looking for it. Modern photographs show no star as bright as magnitude 17 in the position given for it.

There are interesting telescopic objects in Boötes, but not much for the binocular observer. One globular cluster, NGC 5466, is said to be visible, but I have never been able to see it without a telescope. Mu is a star of magnitude 4.5 with a companion of 6.7 at a separation of 109 seconds of arc; the pair is easy enough, though it presents no features of note.

Beta Boötis (3.5) is the nearest fairly bright star to the radiant of the meteor shower seen in early January, known as the Quadrantid shower because this was the region of a constellation, Quadrans (the Quadrant) which was proposed by J. E. Bode in 1775 but which has been deleted from modern maps.

CÆLUM: *the Graving Tool*

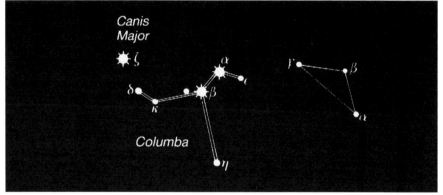

Cælum is one of those constellations which seems to have no claim to separate identity. The brightest stars are Alpha (4.4) and the orange Gamma (4.5), which lies closest to Epsilon Columbæ. Cælum contains absolutely nothing of note.

CAMELOPARDALIS: *The Giraffe*

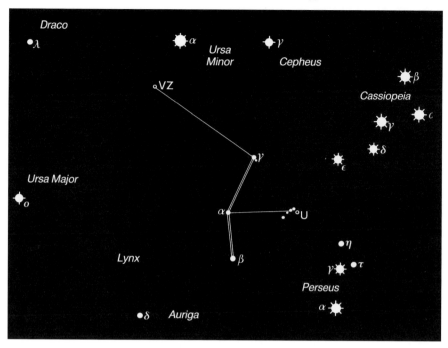

A very barren constellation, adjoining Cassiopeia, sometimes called Camelopardus. Its brightest star, Beta, is only of magnitude 4.0. Two semi-regular variables, U and VZ, can be located with binoculars and are orange in colour, but they have small ranges. Camelopardalis lies between Auriga and the Pole Star.

CANCER: *the Crab*

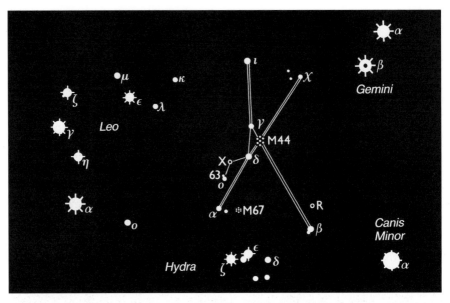

One of the fainter constellations of the Zodiac; it lies between Regulus, in Leo, and Castor and Pollux in Gemini. The only stars above the fourth magnitude are Beta (3.5) and Delta (3.9). The shape of Cancer always reminds me of a very dim and ghostly Orion. Look for the very red N-type semi-regular variable X Cancri, between Delta and the little pair of Omicron and 63; its range is between 5.0 and 7.3, so that it is always easy in binoculars, and is in the same × 7 field with Delta.

Cancer is redeemed by the presence of two fine open clusters, M44 (Præsepe) and M67. Præsepe, nicknamed the Beehive, is one of the finest clusters of its type; it is easily visible with the naked eye and is well seen with × 7 binoculars, while with × 12 and × 20 it is truly glorious. It is flanked by the stars known as the 'Asses'; Delta, which is orange, and Gamma (4.7). M67 is in the field with Alpha (4.2); it is said to be visible with the naked-eye, but I have never been able to confirm this, though with binoculars it is very easy indeed. With × 20, but not with lower magnifications, I can see individual stars in it. M67 has the reputation of being one of the most ancient of the open clusters.

CANES VENATICI: *the Hunting Dogs*

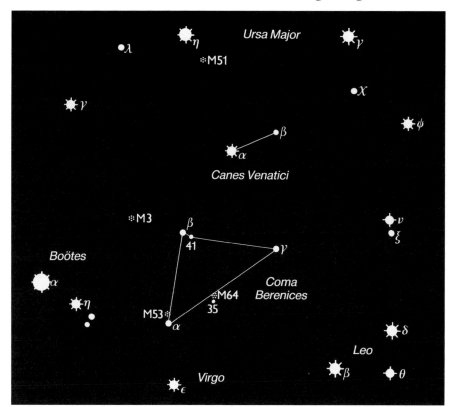

This is a small constellation; the 'dogs' were originally named Asterion and Chara. The only brightish star is Alpha or Cor Caroli (2.9); next comes Beta (4.3) which is in the same × 12 field. Probably the most celebrated object in Canes Venatici is M51, the Whirlpool Galaxy, which is close to Alkaid in the Great Bear. It is of the eighth magnitude, but I admit that I have never been able to see it with certainty even with × 20 binoculars, though it is easy enough in a telescope.

However, it is not difficult to find the globular cluster M3, which has an integrated magnitude of between 6 and 7. It lies at the extreme edge of the constellation, and I have found that the best way to locate it is to use Beta Comæ, which is of magnitude 4.3 and lies rather off a line joining Cor Caroli to Arcturus; though dim, Beta Comæ is rather isolated, and not difficult to identify. In the same × 7 field is a fainter star, 41 Comæ. Swing from 41 through Beta and continue in a direct line until you come to M3. which is just out of the × 7 field with Beta Comæ but which has a fairly obvious star close to it.

M3 is over 48 000 light-years away. With binoculars it looks like a faint blur, easy enough with × 7, while with × 20 there is an indication of resolution into stars at its outer edges. Actually it lies almost midway between Cor Caroli and Arcturus, but I think that the Beta Comæ method of finding it is much easier.

CANIS MAJOR: *the Great Dog*

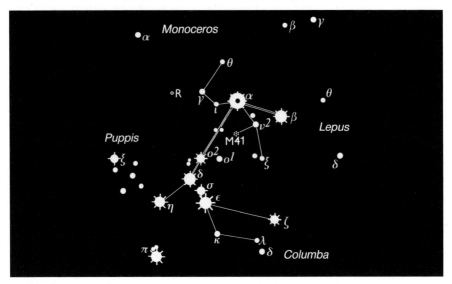

Canis Major, one of Orion's retinue, is graced by the presence of Sirius, the brightest star in the sky. With its magnitude of –1.5 it is well over half a magnitude brighter than its nearest rival, Canopus, and about 1½ magnitudes superior to Alpha Centauri , which is no. 3 on the list. Yet it is far from being exceptionally powerful. It is 'only' 26 times as luminous as the Sun, and owes its pre-eminence to the fact that it is a mere 8½ light-years away from us. Of the really brilliant stars, only Alpha Centauri is closer.

Sirius is pure white, with an A-type spectrum. When low down it seems to flash various colours, because its light is coming to us through a thick layer of the Earth's atmosphere. Even when it is near the zenith, as it can of course be from the southern hemisphere, it still twinkles quite noticeably. Binoculars show it in the guise of a glittering diamond; with × 20 it is almost dazzling.

The other leading stars of Canis Major are Epsilon (1.5), Delta (1.9), Beta (2.0), Eta (2.4), and Zeta and Omicron^2 (both 3.0). Piquantly, all these are far more luminous than Sirius; Delta, over 900 light-years away, could match over 100 000 Suns.

Rich though it is, there are not many binocular objects in Canis Major unless one counts the lovely star-fields near the Milky Way. There is however one splendid open cluster, M4l, which is easily visible with the naked eye and is resolvable with × 20 binoculars; various individual stars can be seen with lower magnifications. It lies in the same field as the rather reddish Nu^2 (3.9), forming a triangle with Nu^2 and Sirius. The brightest star in it is of the seventh magnitude, and is orange, with a K-type spectrum. The colour is obvious with a telescope, but I confess that I have never been able to detect it with binoculars, even × 20. M41 is about 2400 light-years away, and 20 light-years in diameter.

Adjoining Sirius are Iota (4.4), Gamma (4.1) and Theta (also 4.1). Theta is decidedly orange, with a K-type spectrum. I mention it here because it and Sirius show the way to the open cluster M50 in Monoceros.

CANIS MINOR: *the Little Dog*

The junior Dog is a very small constellation, but is unmistakable because of the presence of Procyon, which is one of the nearest of the bright stars; it is 11½ light-years away, and is the equal of 7 Suns. The magnitude is 0.4, so that it is not greatly inferior to Rigel. It is of type F5, and theoretically should be yellowish, but to me it always looks pure white.

The only other conspicuous star is Beta (2.9), which makes up a nice little group with three other stars, Epsilon (5.1), Eta (5.3) and Gamma (4.3). Gamma has a K-type spectrum, and its orange colour is very pronounced with any binoculars. Also in the field is a Mira variable, S Canis Minoris, but as it never rises above the seventh magnitude it is of no real interest to anyone without a telescope.

CAPRICORNUS: *the Sea-goat*

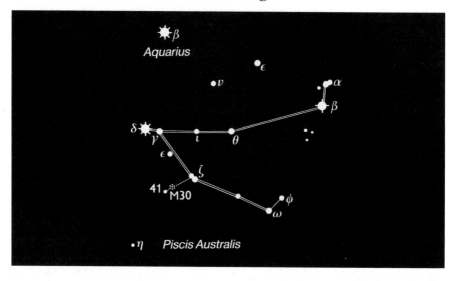

Capricornus is a Zodiacal constellation, but not a brilliant one. The chief stars are Delta (2.9), Beta (3.1), Alpha (3.6) and Gamma and Zeta (each 3.7). There is no really well-defined pattern, but the constellation is not hard to identify, because it lies more or less between Altair and Fomalhaut; the line of three stars of which Altair is the central member points to it.

Alpha is a wide double. The fainter component is of magnitude 4.2, and as the separation is 376 seconds of arc the pair can be split with the naked eye; of course binoculars show it well. Both components are themselves double, though below binocular range. The two main stars are unrelated. The fainter component is about 1600 light-years away (and, incidentally, highly luminous) while the distance of the brighter star is only 117 light-years. Both have spectra of the same type as the Sun. Beta has a sixth magnitude optical companion at 205 seconds of arc. I find it strangely difficult with × 7, but visible with × 8, and easy with × 12 or any higher magnification.

The only other object of immediate interest is the globular cluster M30, near Zeta. It is in the same field as the star 41 Capricorni (5.5) and is close to the limit of visibility with binoculars; I have never seen it clearly without a telescope.

CARINA: *the Keel*

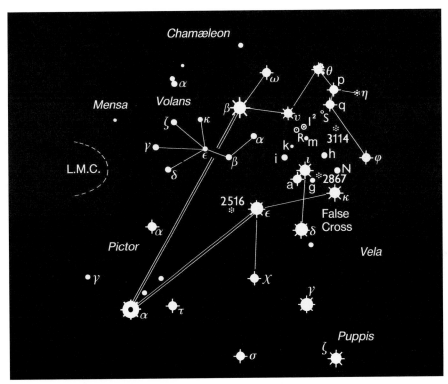

This is the brightest part of the old Argo Navis. Thus its leader Canopus, the brightest star in the night sky apart from Sirus, has become Alpha Carinæ instead of Alpha Argûs.

Canopus can easily be found by using Alnitak and Saiph, in Orion, as direction-finders, but I doubt whether this will be necessary, because Canopus' brilliance makes it stand out at once. The magnitude is –0.7. This is well over half a magnitude fainter than Sirius, but appearances are deceptive; Canopus is a highly luminous star, a very long way away. Estimates of its power and distance vary. The authentic *Sky Catalogue 2000.0* (Cambridge University Press/Sky Publishing Corporation, 1982) gives its luminosity as 200 000 times that of the Sun; even if this is too high, Canopus still qualifies as a cosmic searchlight.

The spectral type is F. In general F-stars are said to be yellowish, but I have never seen any colour in Canopus, either with the naked eye or with optical aid; to me it looks pure white. Its declination is 53 degrees south. This means that it can, in theory, rise from anywhere south of latitude 37 degrees north; it can be seen from Alexandria, but not from Athens, so providing an early proof that the world is not flat! Over parts of Australia and South Africa it sets briefly, but it is circumpolar from Sydney, Cape Town or anywhere in New Zealand.

The other leading stars of Carina are Beta (1.7), Epsilon (1.9), Iota (2.2), Theta (2.8), Upsilon (3.0), Psi and Omega (each 3.3), q and a (each 3.4) and

CARINA

Chi (3.5). The whole constellation is very rich, and is crossed by the Milky Way. Iota and Epsilon make up the famous 'False Cross' with two stars in Vela, Delta (2.0) and Kappa (2.5). There really is some danger of confusion with the Southern Cross; the shape is much the same, and again three of the stars are white while the fourth – in this case Epsilon Carinæ – is an orange giant. However, the False Cross is not only fainter than the Southern Cross, but it is also larger. The four stars are not quite in the same × 7 field. Also, the four members of the False Cross are more equal than those of Crux, where one star in the pattern (Delta Crucis) is markedly fainter than the rest.

The cluster IC 2602 (C102), round Theta, is very fine. Theta itself is bluish white, and about 700 light-years away; in binoculars the whole field is very rich. Probably the best way to locate it is to use Epsilon Crucis, the 'intruder' into the Southern Cross, and Lambda Centauri as direction-finders; Theta Carinæ lies at an almost equal distance on the far side of Lambda Centauri.

Once Theta has been identified, attention can be turned to one of the most intriguing objects in the sky – the strange, erratic variable Eta Carinæ, with its associated nebulosity. Theta and Eta are in the same × 7 field with p and the obviously reddish q, as well as the open cluster NGC 3114, which is not bright, but is easy to see with × 7 and very obvious with any higher magnification. (With × 12, the cluster is in the same field with p and q.)

Eta Carinæ, once brighter than any other star apart from Sirius, is now of about magnitude 7. Binoculars show that it is very highly coloured; telescopically I have described it as an orange blob, quite unlike a normal star. At its peak it was perhaps 6 000 000 times as luminous as the Sun, and this is still true today, though its light is dimmed by intervening nebulosity; in infra-red it is one of the strongest sources in the sky. The Eta Carinæ nebula is visible with the naked eye. Though binoculars will not show the famous dark mass known as the Keyhole, they will bring out the wonderful, varied filaments. It is always worth watching Eta Carinæ; its true nature is uncertain, but there is no reason why it should not flare up again at any time.

NGC 2516 is a lovely open cluster; bright, rather loose, and well seen with × 7. It lies in line with Kappa Velorum and Epsilon Carinæ in the False Cross and is just in the same × 12 field with Epsilon. NGC 2867 (C90), between Iota Carinæ and Kappa Velorum, is a planetary nebula. It is rather faint; I can just see it with × 12 and easily with × 20, but I have never been able to detect it

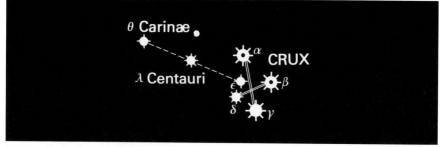

How to find Theta Carinæ (C102).

with × 7. Using × 7, Iota is in the same field with g (4.3, orange), l (a Cepheid, with a range of from 3.4 to 4.8 and the unusually long period, for a Cepheid, of 35 days), R (a Mira variable, with a range of from 4 to 10 and therefore easy with binoculars for part of its 309-day period) and the red N Velorum (3.1). R Carinæ is just in the × 12 field with Iota. Even when it is near maximum, I find the colour much less pronounced than with most other Mira stars.

Eta Carinæ photographed by D.F. Malin (Anglo Australian Telescope Board, © 1977).

CASSIOPEIA: *Cassiopeia*

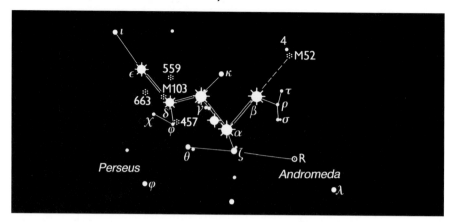

Cassiopeia, one of the most distinctive constellations in the northern hemisphere of the sky, is circumpolar from England, northern Europe and the northern United States. It cannot possibly be mistaken. In case of any doubt, locate it by a line from Mizar in the Great Bear through Polaris and prolonged; Cassiopeia lies at about an equal distance beyond Polaris, so that when the Bear is low down Cassiopeia is high up, and vice versa. Cassiopeia is crossed by the Milky Way, and the five leading stars make up the familiar W or M pattern: Gamma (2.2, though variable), Alpha (also 2.2, slightly variable), Beta (2.3), Delta (2.7) and Epsilon (3.4).

Alpha or Shedir is an orange K-type star; the colour is very evident in binoculars. Having observed it for over 50 years I am convinced that it is variable over a small range, perhaps from 2.0 to 2.4, but the fluctuations are very slow, and some catalogues give the magnitude as constant. Beta makes a good comparison star. Gamma is quite definitely variable, and is an unstable star, with a luminosity about 6000 times that of the Sun; periodically it throws off shells of material, though these cannot be detected visually. I first noticed its behaviour (independently!) in 1936, when I was making an estimate of the brightness of Shedir, and found that Gamma had brightened appreciably from its catalogue magnitude of 2.2. In early April 1937 it rose to 1.6, so that it was considerably brighter than Polaris. It then declined, dropping to below the third magnitude in 1940. Subsequently it increased again, and for several decades now the magnitude has hovered around 2.2. It is always worth watching, because its behaviour is quite unpredictable.

Even more puzzling is Rho Cassiopeiæ, if only because nobody knows what sort of a variable it is. It is easy to find, because it is usually visible with the naked eye at about magnitude 5, and is in the same binocular field with Beta; to either side of it are Tau (5.1) and Sigma (4.9), which act as good comparisons, though Tau is of type K and has been suspected of slight variability. Rho has a spectrum of type F8, and is slightly 'off white'; I have even recorded it as perceptibly orange, which is unusual for an F-star. Its distance and luminosity are wildly uncertain, and estimates of its power vary. It may be at least 200 000 times as luminous as the Sun. Ever since I started observing it, about 40 years ago, the magnitude has remained between 4.6 and 5.3, with a mean of 4.9, but there are occasions when it has fallen to below 6. It has been classed as an R Coronæ star, but it is certainly atypical, and it is wiser to admit that at present we simply do not know. It should be on every binocular observer's list.

The red Mira variable R Cassiopeiæ lies some way from the W shape, between Shedir and Lambda Andromedæ. It has a period of 431 days, and a range of between 5.5 and 13, so that for most of the time it is well below binocular range. It is not easy to identify, because it lies in a rich area and there are no convenient guides to it.

There are several clusters in Cassiopeia. The open cluster M52 lies in line with Shedir and Beta; to find it, first identify the star 4 Cassiopeiæ, of magnitude 5.2, and you will find M52 in the same field with × 20. It is clear enough, with slight resolution into stars. It is just in the Beta field with × 7, just out of it with × 12. With × 7 I find it rather difficult to see. The diameter is about 15 light-years, and the leading stars in it are hot and white, so that, like the Pleiades, the cluster is fairly young by cosmical standards.

NGC 663 (C10) is a prominent open cluster in the same × 8.5 field with Delta and Epsilon; look for it just off a line joining these two stars. It is very well seen with × 20. Between it and Delta is another open cluster, M103, but it is very poor and sparse, and it is hard to understand why Messier listed it and omitted NGC 663. M103 is in the same × 20 field with Delta, but is not at all easy to identify, and looks rather more like a chance aggregation of stars.

Yet another inconspicuous open cluster is NGC 559 (C8), which makes a triangle with Delta and Epsilon, roughly between Epsilon and Gamma. I always find it a difficult binocular object; I can just see it with × 20, but I am not confident that I can identify it with a lower magnification.

NGC 457 (C13) is a far more interesting cluster. It is easy to find from the stars Chi (4.7) in Phi (5.0), which make up a small triangle with Delta and are in the same × 20 field with it. The distance of the cluster is believed to be over 9000 light-years; it contains several thousands of stars, and it cannot be less than 30 light-years in diameter.

The unusual factor is that Phi Cassiopeiæ lies on the south-east edge of the cluster. Whether or not it is a genuine member remains doubtful. If it is. then it must be well over 200 000 times as luminous as the Sun, much more powerful than Rigel or Deneb.

With × 7, the cluster appears to me only as an extremely dim blur adjoining Phi. With × 12 the haze is much more pronounced, and an unwary observer could easily mistake it for a comet. With × 20 it is easy, though I have not been able to obtain any definite resolution into stars without using a telescope. Altogether it is an exceptional object, and I know of nothing else quite like it within the range of binoculars.

CENTAURUS: *the Centaur*

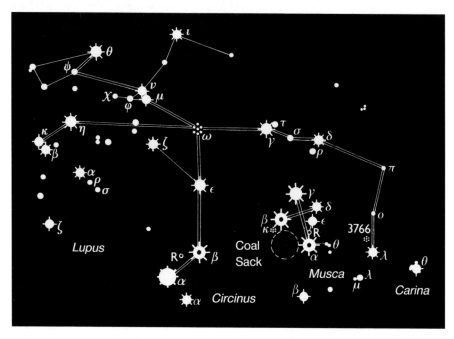

This is one of the largest and most important of all the constellations. The leading stars are Alpha (−0.3), Beta or Agena (0.6), Theta (2.1), Gamma (2.2), Epsilon (2.3), Eta (2.3, but very slightly variable), Zeta (2.5), Delta (2.6), Iota (2.8), Mu (3.0), Kappa and Lambda (each 3.1) and Nu (3.4). Of these, the most northerly is the orange K-type Theta, with a declination of 36 degrees south; therefore it rises anywhere south of latitude 54 degrees north, whereas the two Pointers to the Southern Cross, Alpha and Agena, cannot be seen from any part of Europe. From Sydney or Cape Town, the Pointers never set.

With × 7, the Pointers are in the same field. They are different in colour; Alpha is somewhat yellowish, while Agena is bluish-white. Alpha, of course, is the nearest of all the bright stars, at a mere 4.3 light-years; it is a fine binary with a period of 80 years, but the components are too close to be split with binoculars. Strangely enough Alpha Centauri has no official proper name, though it has been called 'Rigel Kent' or 'Toliman'. Agena, at 460 light-years, is over 10 000 times as luminous as the Sun. Looking at the Pointers, apparently side by side in the sky, it is not easy to appreciate that they are quite unconnected with each other.

Omega Centauri (C80) is the brightest of all the globular clusters. It is easily visible with the naked eye, and is striking in binoculars; with × 12 there is some resolution of its outer parts, and with × 20 it is seen to be starry. To find it, simply follow a line from Beta through Epsilon; Omega lies at an equal distance on the far side of Epsilon. With × 7, though not with × 12, Epsilon and Omega are just in the same field. Omega is around 17 000 light-years away; it may contain more than a million stars. The average separation between the stars at its centre is no more than a tenth of a light-year.

Omega Centauri (C80), brightest of all globular clusters (© Jack Newton).

A different kind of cluster, NGC 3766 (C75), is in the same field as Lambda, and just in the × 7 field with Alpha Crucis. It is an open cluster, easily seen with × 7; with × 12 it seems to me to have a slightly elliptical appearance, with indications of resolution into stars. Beta and Gamma Crucis point to Delta Centauri, the centre of a fine binocular field which also contains Rho (4.0). Look also for R Centauri, a red Mira-type variable more or less between the Pointers. At maximum it rises to 5.4, and is an easy binocular object, though at minimum it sinks to below magnitude 11. It has the rather unusually long period of 547 days.

The whole of Centaurus is very rich, and well worth sweeping with a low magnification.

CEPHEUS: *Cepheus*

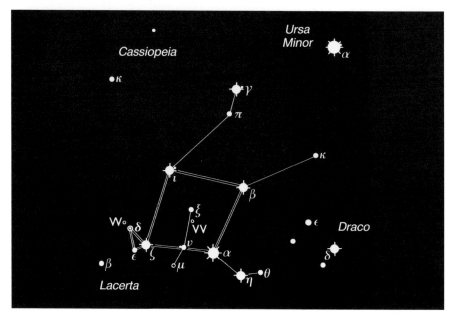

Cepheus, in the far north of the sky, contains no very bright stars; its leaders are Alpha (2.4), Gamma and Beta (each 3.2) and Zeta (3.3). Cepheus lies inside the triangle outlined by Deneb, Polaris and Cassiopeia. There is no really distinctive shape, but there are two objects of special interest.

One is Delta Cephei, which has given its name to the whole class of short-period variables. The range is from 3.5 to 4.4. and the period is 5.4 days. Delta is one of a small triangle; the other members are Zeta, which is a K-type star showing as clear orange in binoculars, and Epsilon (4.2). These make good comparisons, though Delta never becomes quite as bright as Zeta. With × 8.5, the three are in the same field.

The other famous variable is Mu, William Herschel's so-called Garnet Star. It is without doubt the reddest of all the naked-eye stars, though binoculars are needed to bring out its colour properly. The range is between 3.6 and 5.1, and there seems to be no semblance of a period. It is in the same binocular field with Nu (4.3), and is usually comparable with it. Using × 8 or higher, Mu looks like a glowing coal. It is very luminous – at least 100 000 times as powerful as the Sun, and therefore much superior to Betelgeux, though it is much further away (over 1500 light-years). Incidentally Nu, which looks so ordinary, is also very luminous and remote.

Xi Cephei (4.3) is just out of the × 7 field with Nu. Close beside Xi is the huge eclipsing binary VV Cephei, which has the very long period of 7430 days; the next eclipse is not due until 2010, when the magnitude will drop from its usual 4.9 to about 5.2. The main component is of type M, and is one of the largest stars known. Binoculars bring out its decidedly orange colour. A third variable in the constellation is W Cephei, with a range of from 6.9 to 8.6. The changes are very slow; the period is thought to be rather over 1000

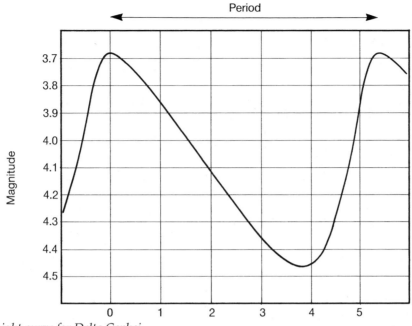

Period

Magnitude

Light curve for Delta Cephei.

days, but is probably not constant. The positions of the variables are shown on the additional chart.

On the whole Cepheus is rather a barren group, but the presence of Delta and Mu redeems it from the viewpoint of the binocular observer.

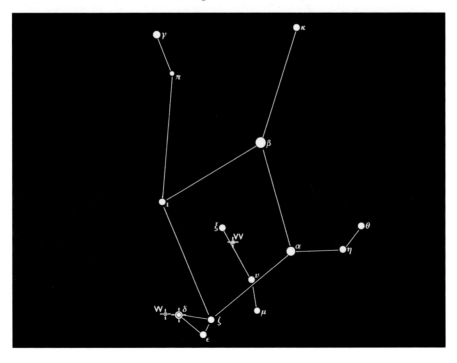

CETUS: *the Whale*

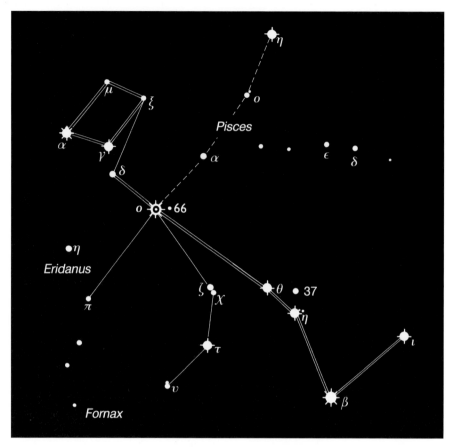

Cetus is one of the largest constellations in the sky, but by no means one of the richest. It has only two stars above the third magnitude, Beta or Diphda (2.0) and Alpha (2.5); next come Eta (3.4) and Gamma and Tau (each 3.5). The best way to find Diphda is to use Alpheratz and Gamma Pegasi, in the Square of Pegasus, as pointers. Diphda is of type K, and is obviously orange; it has been suspected of variability, but is awkward to estimate, because there are no suitable comparison stars anywhere near it. It is 68 light-years away, and 60 times as luminous as the Sun.

Do not confuse Diphda with Fomalhaut in the Southern Fish. There should be no real problem, because Diphda is considerably the fainter of the two.

The 'head' of Cetus is made up of Alpha, Gamma, and Mu and Xi (each 4.3). Alpha is an M-type giant, 120 times as luminous as the Sun. It is a fine sight in binoculars, and the orange colour is noticeable even with the naked eye.

Theta (3.6) and Eta make up a pair. Theta is white, and Eta, with a K-type spectrum, slightly orange. In the same × 12 field you will find 37 Ceti, which is double; the components are of magnitudes 5 and 7.8. and the separation is 50 seconds of arc. I can just see both components with × 20, but there is nothing remarkable about them.

Tau Ceti is a close neighbour, at a distance of only 12 light-years. It has been regarded as one of the two nearby stars sufficiently like the Sun to be candidates as centres of planetary systems (Epsilon Eridani is the other). Tau Ceti is rather smaller, dimmer and less luminous than the Sun.

Much the most celebrated object in the constellation is Omicron Ceti or Mira, the prototype long-period variable. It was identified as variable by the Dutch astronomer Phocylides Holwarda as long ago as 1638, since when almost every maximum has been observed. The mean period is 331 days, but both the period and the amplitude change from one cycle to another. At some maxima Mira can reach the second magnitude, and in 1779 it is said to have risen to 1.7, much brighter than Polaris, while in 1969, 1977 and 1987 I estimated it as 2.3 at its peak, though on other occasions it never exceeds magnitude 4. At minimum it sinks to 10, well out of the range of binoculars. It is a large star, perhaps as much as 40 000 000 kilometres in diameter, but it is comparatively cool; it is of course red, though the colour is most pronounced at minimum.

Mira lies in line with Eta, Omicron and Alpha Piscium. It is just in the × 7 field with Delta Ceti (4.1) and not far from 66 Ceti, which is a telescopic double. My advice is to locate it when it is bright, and then memorize its position so that you can find it again when it has faded. Because its period is not much more than a month shorter than a year; there are times when maxima occur with Mira too close to the Sun to be seen.

CHAMÆLEON: *the Chameleon*

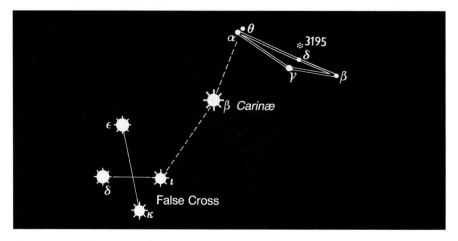

A small far-southern constellation. Probably the best way to find it is to extend a line from Iota Carinæ, in the False Cross, through Beta Carinæ and prolong it for some distance.

Alpha, Beta, Gamma and Delta are all between magnitudes 4.0 and 4.5. They form a rough diamond shape, and are in the same × 12 field, so that they are easy enough to recognize in spite of their dimness. Theta (4.3) is in the same field with Alpha (4.1), making up what looks like a wide pair; Alpha is white, Theta decidedly orange. Delta is made up of two components 6 minutes of arc apart; the brighter member is white, while the fainter is orange. The colour contrast is marked enough to be noticeable even with × 7. Gamma (4.1) is another orange star.

NGC 3195 (C109) is a globular cluster in the field with Delta. Its magnitude is officially given as below 9, so that it should be invisible with binoculars; I have suspected it with × 12 or higher magnification, but it is difficult to be sure, as there are scattered disconnected stars around. Chamæleon contains nothing else of note.

CIRCINUS: *the Compasses*

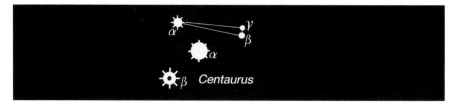

A small constellation with only one fairly bright star. Alpha (3.2) which is in the same × 7 field with Alpha Centauri. It contains nothing of interest to the binocular observer.

COLUMBA: *the Dove*

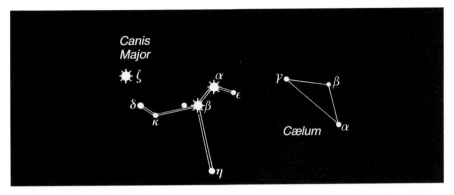

Originally Columba Noachi, Noah's Dove. The brightest stars are Alpha (2.6), Beta (3.1), Delta (3.8) and Epsilon (3.9); Beta and Epsilon are orange, with K-type spectra. and so is Eta (4.0). Columba may be identified by the curved line of stars south of Lepus. It is always very low from England, and Eta does not rise at all. For southern observers, a good way to find it is to join Canopus to Rigel by an imaginary line; this will cross Columba and Lepus.

COMA BERENICES: *Berenice's Hair*

At first glance this constellation gives the impression of being a vast, faint open cluster. The brightest stars (Beta, Alpha and Gamma) are only of magnitude 4.3, but the area is very rich, and well worth sweeping, particularly with low magnification (I find that × 7 is best). The globular cluster M53 is in the same field as Alpha. It is above the eighth magnitude, and is visible with × 7, though only as a dim blur; its real distance is 69 000 light-years, so that it is much further away than the brighter globular clusters such as M13 Herculis and Omega Centauri. With × 12 it is easy, and with × 20 quite distinctive, though I have never been able to resolve even its outer parts without using a telescope. Alpha Comæ forms a triangle with Eta Boötis and Epsilon Virginis, and is not hard to identify, as there are no other naked-eye stars close to it.

Coma is crowded with galaxies. Most of these are well beyond binocular range but it is worth looking for the 'Black-Eye' galaxy M64, which is within a degree of the star 35 Comæ, not far from Alpha. Its magnitude is given as 6.6, but I have always found it an elusive binocular object.

Beta Comæ and its neighbour 41 act as good guides to the globular cluster M3, which lies just across the border of Canes Venatici and is described under that heading.

CORONA AUSTRALIS: *the Southern Crown*

Though small and faint, this little constellation is easy to find, close to Alpha Sagittarii. Its brightest stars are Alpha and Beta (each 4.1); with the rather fainter Gamma, Delta and Theta the little semicircle is characteristic enough. NGC 6541 (C78), between Theta Coronæ and Theta Scorpii, is a globular cluster just detectable in binoculars. I have included Corona Australis in the map with Sagittarius.

CORONA BOREALIS: *the Northern Crown*

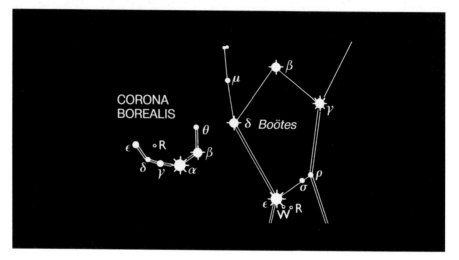

This is very easy to find. I have also included it with the map of Boötes, since it so obviously belongs to the Boötes pattern. The leading star is Alpha or Alphekka (2.2). The semicircle made up of Epsilon (4.1), Gamma (3.8), Alpha, Beta (3.7) and Theta (4.1) is unmistakable. Epsilon, of type K, is 'off-white' in binoculars.

It is interesting to see how many stars you can count inside the 'bowl' of Corona. 7 × 50 binoculars show at least 15.

R Coronæ, in the bowl, is generally well within binocular range, and can just become visible with the naked eye, but when it drops to minimum it may fall to as low as magnitude 15. Fortunately there is a good comparison star of magnitude 6.6, also in the bowl. If you look at Corona and find only one star in the bowl above the seventh magnitude, you may be sure that R Coronæ has faded.

The other object of special note is the so-called Blaze Star, T Coronæ, which is usually of about magnitude 10, but which flared up to naked-eye visibility in 1866 and again in 1946. It is not likely to have another outburst yet awhile, but it is certainly worth watching; with recurrent novæ, one never knows. It lies in the same × 7 field with Epsilon.

CORVUS: *the Crow*

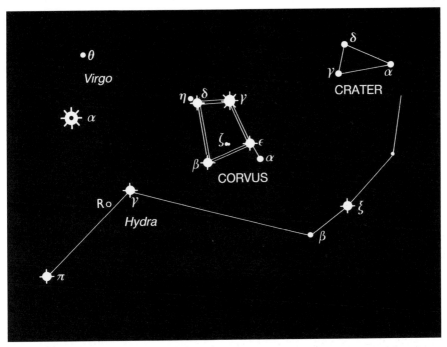

Corvus adjoins Hydra, and is easily recognizable because of the four leading stars, which make up a quadrilateral: Gamma and Beta (each 2.6), and Delta and Epsilon (each 3.0). Epsilon has a K-type spectrum, and is slightly orange. Delta is in the same field with Eta (4.3). Delta and Epsilon are in the same field with Zeta (5.3), which is one of a pair of faint stars. There is little of interest in Corvus; curiously, the star lettered Alpha is only of magnitude 4.0.

CRATER: *the Cup*

Like Corvus. a small constellation adjoining Hydra. Its brightest star is the slightly yellowish Delta (3.6), which makes up an inconspicuous triangle with Gamma and Alpha (each 4.1). The nearest fairly bright star to the constellation is Nu Hydræ.

CRUX AUSTRALIS: *the Southern Cross*

The most famous of all the southern constellations was only separated from Centaurus in 1679. It is more or less surrounded by Centaurus; the only other common boundary is with Musca. The chief stars are Alpha or Acrux (0.8), Beta (1.2), Gamma (1.6), Delta (2.8) and Epsilon (3.6), which rather spoils the symmetry of the pattern. For obvious reasons, I have included Crux in the same map with Centaurus.

The four main stars, making up the kite-pattern, are in the same × 7 field and just fill the field with × 12. Acrux is a wide double, though not separable with binoculars. In the same × 7 field it is easy to find the very wide pair of Theta^1 (4.3) and Theta^2 (4.7), which are not genuinely associated; as noted earlier, the dimmer component is much the more distant of the two.

Even a casual glance will show that while three of the main stars are white, the fourth – Gamma – is warm orange; it is an M-type giant, and the colour contrast is striking. The Cepheid variable R Crucis, between Acrux and Epsilon, is an easy binocular object, though it never becomes much brighter than the seventh magnitude. Both Mu (4.0) and Iota (4.7) are double, but I cannot separate them even with x 20.

The glorious cluster NGC 4755 (C94) is often nicknamed the Jewel Box, though officially known as Kappa Crucis. It is impossible to overlook; it adjoins Beta, and is in the same binocular field. Most of its stars are bluish-white, but there is one red supergiant which stands out; I can just about detect its colour with × 12, though admittedly a telescope is needed to show the cluster in its full splendour. It is around 7700 light-years away, with a central region about 25 light-years in diameter. It is relatively young, with a probable age of no more than a few million years. Three of its leading stars make up a triangle, inside which lies the red supergiant.

By sheer coincidence Kappa Crucis lies at the edge of the dark nebula known as the Coal Sack (C99). There is no real connection; the Coal Sack is no more than 500 light-years away, so that it is in the foreground. With binoculars it appears as a virtually blank area, about 7 degrees by 5 degrees in extent; there are a few faint stars in front of it.

Crux is useful as a guide. With its declination of 63 degrees south, Acrux is more or less circumpolar from much of South Africa and Australia, and never sets over New Zealand. It does not rise over La Palma in the Canary Islands, where the new observatory has been set up, but is easy enough from Hawaii.

CYGNUS: *the Swan*

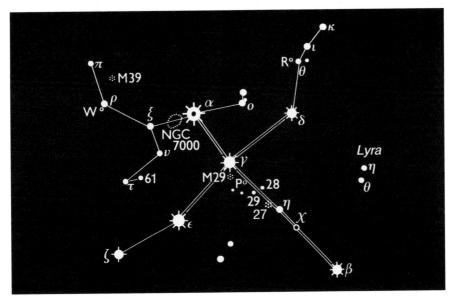

Cygnus is one of the most impressive constellations in the sky. It is often nicknamed the Northern Cross, and certainly it is much more cruciform than the Southern Cross. The leading stars are Alpha or Deneb (1.2), Gamma (2.2), Epsilon (2.5), Delta (2.9) and Beta or Albireo (3.1) which make up the cross; next in order of brightness is Zeta (3.2).

Deneb is exceptionally luminous, and is the equal of at least 60 000 Suns. It is 1800 light-years away, so that we see it today as it used to be in Roman times; it is white, with an A-type spectrum. Albireo is the faintest of the stars in the X, and is further away from the centre of the pattern, marked by Gamma; but to compensate for this, Albireo is the loveliest double in the sky, with a golden-yellow primary and a 5.4-magnitude blue companion. I can just about split it with × 20, but not easily.

Close to Deneb is the little group made up of Omicron[1] (3.8), Omicron[2] (4.0) and 32, which make up a trio in the same × 12 field. The intensely red Mira variable U Cygni is in the same field with 32. It can attain magnitude 6.7, and is then within binocular range, but for most of its 465-day period it is too faint. Another Mira variable is R Cygni, in the same field with Theta (4.5); Theta is in line with Iota (3.8) and Kappa (also 3.8), and can be identified by the fainter star close beside it. At its best, R Cygni can reach magnitude 6.5, but at minimum it sinks to below 14. When within binocular range it may be seen on the side of Theta opposite to the 6.5-magnitude star. The period is 426 days, and the spectral type is S.

More prominent, at least at times, is Chi Cygni, not far from the reddish Eta (3.9), roughly between Gamma and Albireo. Here we have another S-type Mira variable, with a period of 407 days. The range is the largest known. At some maxima it can reach magnitude 3.6, outshining Eta, though generally the peak magnitude is below 4; at minimum it becomes very faint – below

85

magnitude 14. (At infra-red wavelengths, Chi Cygni is one of the brightest objects in the entire sky.) It was the third variable star to be identified, by Kirch as long ago as 1686.

Chi is in the same × 12 field with Eta. When near maximum it cannot be mistaken, and it is of course very red, but do not be surprised if you fail to find it, because for most of its period it is well below binocular range. Yet another variable, this time a semi-regular, is W Cygni, in the same field with Rho (4.0). The range is from 5.0 to 7.6, so that the star is never too dim to be found with binoculars; there is a very rough period of around 130 days. Even with x 7 the colour is noticeable, and is very marked with × 12 or higher power.

P Cygni is different. In 1600 it flared up from obscurity to the third magnitude; ever since 1715 it has fluctuated around 5, and may be compared with 28 Cygni (4.8) and 29 Cygni (5.0). It is highly luminous, and is unstable, periodically throwing off shells of materials. The distance has been given as 4500 light-years, but may be as much as 7000, in which case it must be more than a million times as luminous as the Sun. It is unlikely to produce any spectacular outbursts in the near future, but one never knows.

The Milky Way flows through Cygnus, and this is one of the richest areas in the sky, so that it is well worth sweeping; note the dark rifts which indicate the presence of obscuring material.

There are several notable clusters and nebulæ. The open cluster M29 is in the 'bowl' which contains P Cygni, and is in the same × 12 field with both P and Gamma, which makes it easy to locate. With low powers it appears as a dim blur; I have suspected some resolution with × 20, but not with any real confidence. It is around 7000 light-years away, and would appear much brighter were it not dimmed by interstellar material between the cluster and ourselves. In the same × 7 field, outside the 'bowl' in the direction of Eta, is 27 Cygni (NGC 6871), a very sparse, loose cluster which hardly looks like a cluster at all with low magnifications, though with × 20 the grouping is fairly evident.

Another open cluster, M39 is in the × 7 field with Rho (4.0) and Pi (4.7); the × 12 field will just contain the three objects. M39 is very loose, but sufficiently condensed to make it obvious, particularly in view of the characteristic pattern of its leading stars. Its distance is of the order of 800 light-years; altogether it contains some 30 stars.

The constellation also contains an object of quite different type: NGC 7000 (C20), the so-called North America Nebula. When photographed with adequate equipment it really does recall the outline of the North American continent; it is dimly visible with the naked eye in the guise of a slightly brighter section of the Milky Way, and binoculars show it clearly as a large region of diffuse nebulosity. With × 7 and × 8.5 its shape is not readily identifiable, but with × 12 the outline becomes more definite, and with × 20 the North American shape can be distinctly made out. The nebula is nearly 50 light-years in diameter, and may owe much of its illumination to Deneb.

To locate it, first find Deneb and the reddish Xi Cygni (3.7). With × 7 or × 8.5 the Nebula is then in the same field. With × 12 the field will barely cover all three, but once the Nebula has been found there should be little difficulty in finding it with × 20. It is certainly a wonderful sight.

DELPHINUS: *the Dolphin*

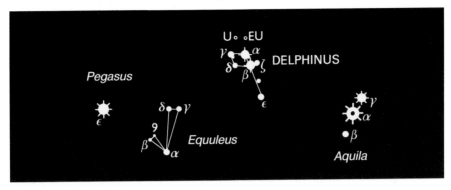

There can be no mistaking Delphinus. It looks almost like a cluster, and I have known unwary observers to confuse it with the Pleiades. It adjoins Aquila, and may be found by sweeping away from the line of stars of which Altair is the central member.

The chief stars of Delphinus are Beta or Rotanev (3.5), Alpha or Svalocin (3.8), Gamma (3.9), Epsilon (4.0) and Delta (4.4). (The curious names of Svalocin and Rotanev are due to the fact that they were christened by an otherwise obscure astronomer named Nicolaus Venator!) All the stars of the pattern are in the same field with × 12 or lower. Alpha has a fainter star beside it, giving the misleading impression of a very wide double. Do not confuse it with the Beta–Zeta pair, which is more widely separated; Zeta is of magnitude 4.7.

The most interesting objects are the red variables U and EU, which are easy to locate. U, of spectral type M, appears to be irregular, with a range of from 5.6 to 7.5; EU ranges from 6.0 to 6.9, and is also of type M. It is said to be semi-regular, with a period of around 60 days, but I have never been at all confident about this. There are two good comparison stars shown, in the special chart.

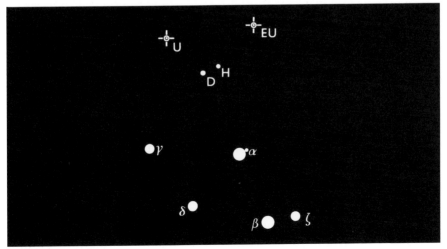

U and EU Delphini.

DORADO: *the Swordfish*

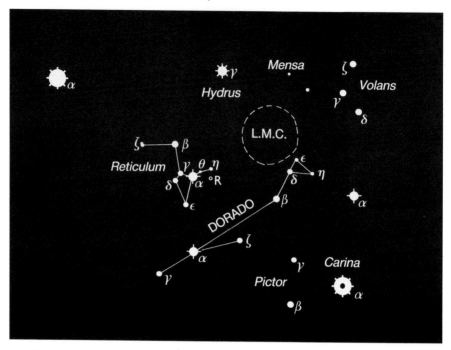

Dorado is in the far south of the sky, adjoining Reticulum. Its leading star, Alpha (3.2), lies roughly between Canopus and Achernar. Beta is a Cepheid variable, one of the brightest members of its class; the range is from 3.8 to 4.7, and the period is 9.8 days. Suitable comparison stars are Delta (4.3), Zeta (4.7) and Eta (4.9). The red semi-regular variable R Doradûs lies near Alpha, and since its range is from magnitude 7 to 8 if can be followed with binoculars.

Dorado contains most of the Large Cloud of Magellan (a small part of the Cloud extends into Mensa). Any binoculars will give a superb view of this huge system, which is much the brightest of the really large external galaxies. The outline is very obvious. Look in particular for the complicated Tarantula Nebula (30 Doradûs), which is the most brilliant part of the whole Cloud and can be identified even with the naked eye; it is the largest known diffuse nebula, and if it were as close to us as M42 in Orion it would cast strong shadows. Use as high a magnification as possible, though it is true that the real complexity of the nebula cannot be properly seen without a telescope.

DRACO: *the Dragon*

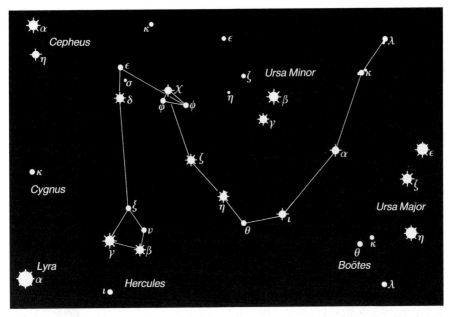

This long, sprawling constellation extends from near Vega through to the gap separating the two Bears. It therefore adjoins Ursa Major, and in ancient times Thuban or Alpha Draconis used to be the north polar star. Thuban lies more or less between Mizar and Beta Ursæ Minoris (Kocab), but is only of magnitude 3.6. The leading stars of Draco are Gamma (2.2), Eta (2.7), Beta (2.8), Delta (3.1), Zeta (3.2) and Iota (3.3). Gamma, a decidedly orange K-type star, makes up the Dragon's head with Beta, Xi (3.7) and Nu, which is a wide double; each component is of magnitude 4.9, and the 'twins' are identical. Keen-sighted people can distinguish them both with the naked eye. I am not sure that I can do so, but with even × 7 binoculars they are clear enough. Both stars are white. They make up a physically-associated system, but the real separation between them is over 300 000 million kilometres. With × 7, all four stars in the Dragon's head are in the same field.

To the binocular-owner there is little else of interest in Draco, which is a decidedly barren group. However, it is worth noting that Sigma (4.7), between Delta and Epsilon and in the same field with Epsilon is one of the nearer stars, only 19 light-years away and only about a third as luminous as the Sun. Unusually for so dim a star, it is dignified by an old proper name: Alrakis.

ERIDANUS: *the River*

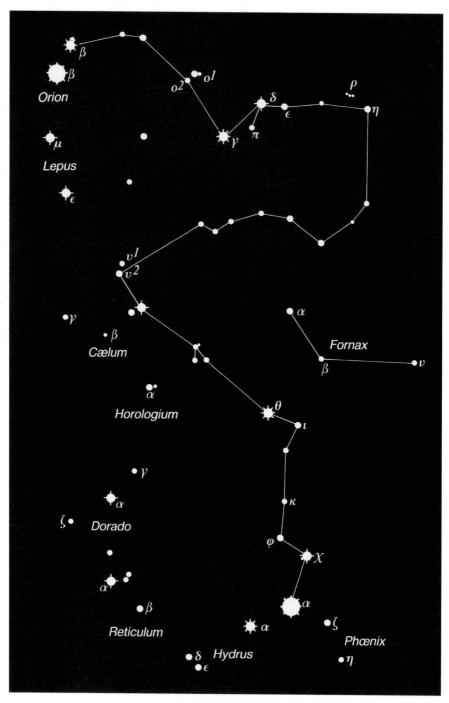

An immensely long, winding constellation, extending from close to the equator, near Rigel in Orion, through to the south polar region. Achernar (0.5) is one of the most brilliant stars in the sky; then come Beta (2.8), Theta and Gamma (each 2.9), Delta (3.5) and Upsilon[4] and Psi (each 3.6).

Only part of Eridanus is visible from Europe and North America. Achernar, the closest really bright star to the south pole, can be seen from Cairo or anywhere in a more southerly latitude; from New Zealand it is circumpolar. It is bluish-white, and 780 times as luminous as the Sun; its distance is 85 light-years. At the other end of the constellation is Beta, near Rigel and only 5 degrees south of the celestial equator.

There is some mystery surrounding Theta Eridani, or Acamar. Ptolemy ranked it as of the first magnitude, and seems to have referred to it as 'the Last of the River'; from Alexandria he could not see Achernar, though Acamar was visible low over the horizon. Either there has been some mistake, or else Acamar has faded. Probably there has been no real change. Acamar is a superb telescopic double, but as both components are of spectral type A they are not likely to have become dimmer in historical times.

It is worth looking at the Delta–Epsilon pair, close to Gamma. Delta is of type K, and with $\times 7$ or higher magnification is very clearly orange; it is in the same field as Pi (4.4) which is a red star of class M. Epsilon (3.7) is one of the very closest stars, at 10.7 light-years. It is smaller and cooler than the Sun, but is not too unlike it, and has often been regarded as a possible planetary centre. In 1983 the Infra-Red Astronomical Satellite, IRAS, discovered that it is associated with cool material which may indeed be planet-forming. though as yet there is no proof. Like Delta, Epsilon has a K-type spectrum, but I find its orange hue much less pronounced.

Forming a triangle with Gamma is the little pair of Omicron[1] (4.0) and Omicron[2] (4.4). Omicron[2] is a remarkable triple system, one member of which is a White Dwarf. Telescopes are needed to show the White Dwarf, but binoculars show the main pair clearly, and the components arc separable with the naked eye.

Despite its huge area, there is not much else of immediate interest in Eridanus. Note however the little trio of Rho[1], Rho[2] and Rho[3], in the same field with Eta (3.9).

EQUULEUS: *the Foal or Little Horse*

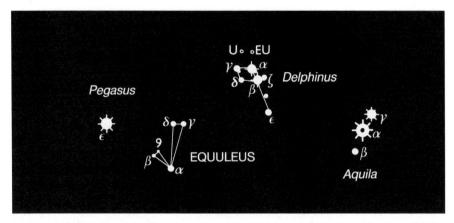

This is one of the most obscure of Ptolemy's original constellations. To find it, use Alpha and Delta Delphini as pointers. With × 7, Beta and Delta Equulei are just in the field with Epsilon Pegasi.

The chief stars, Alpha (3.9), Delta (4.9) and Gamma (4.7) form a triangle; do not confuse it with the smaller, fainter triangle formed by Alpha, Beta and 9 Equulei. Gamma and 6 Equulei are close together, and look like a wide double, though they are not actually associated with each other. I have included Equuleus in the map with Delphinus.

FORNAX: *the Furnace*

Fornax contains only one star above the fourth magnitude, Alpha (3.9); the best guide to it is the curved line of stars in Eridanus lettered Tau^2 to Tau^6. Fornax is always very low from European latitudes; it contains an important cluster of galaxies, but all its members are below binocular range. I have included Fornax in the map with Eridanus.

GEMINI: *the Twins*

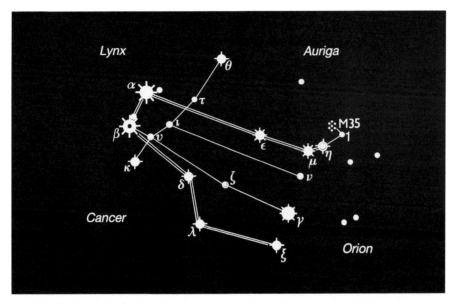

The Heavenly Twins, Castor and Pollux, dominate this important constellation. They are not alike. Pollux is a K-type star, whose orange colour is noticeable with the naked eye and unmistakable in binoculars; Castor is white, and is an interesting multiple system, though binoculars show it as single. Pollux (1.1) is much brighter than Castor (1.6), though ancient astronomers described it as fainter. The two are not associated, since Castor is 53 light-years away and Pollux only 36. The other leading stars in Gemini are Gamma or Alhena (1.9), Mu (2.9), Epsilon (3.0), Eta (3.0 at maximum), Xi (3.4) and Delta (3.5). The constellation has a distinctive shape, and there are lines of stars extending from Castor and Pollux in the general direction of Betelgeux.

There are two notable variables here. One is Eta, a semi-regular star with an official range from magnitude 3 to 3.9 and a period of around 233 days. The situation is complicated by the presence of an invisible binary companion which can pass in front of the bright star, so that the total magnitude falls. The obvious comparison star is its neighbour, Mu. Both it and Eta have M-type spectra, and are obviously orange when seen through binoculars; it is interesting to compare their colours, and I have often found Eta the more strongly coloured of the two, though the difference is very marginal. In the same field lies 1 Geminorum, not much use as a comparison for Eta because it is too faint (4.1).

The other bright variable is Zeta, a typical Cepheid with a range of from 3.7 to 4.3 and a period of 10.2 days. Useful comparison stars are Delta (3.5), Lambda (3.6), Kappa (3.8) and Nu (4.1); binoculars show that Nu is slightly orange. With × 7 Zeta is in the same field as Delta and Lambda, and this makes comparisons easier even though Zeta is always visible with the naked eye under even reasonable conditions of seeing.

GEMINI

The most spectacular binocular object in Gemini is the open cluster M35. It is visible with the naked eye, and its position close to Eta and Mu makes it very easy to locate. This is one of the brightest of all open clusters. With × 7 it appears as a filmy mass; I suspect stars with × 8.5, and with × 20 there are plenty of stars to be seen in it, though binoculars do not bring out any distinctive shape. It was recorded as early as 1749, by the French astronomer Legentil, though presumably it must have been noticed in more ancient times.

The Milky Way passes through Gemini, near Mu and Gamma. This is a very rich region, and will repay sweeping with a low power.

M35, open cluster in Gemini (© Jack Newton).

GRUS: *the Crane*

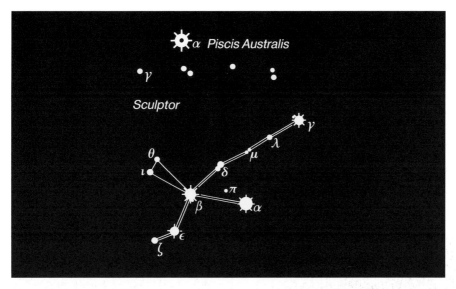

Grus is much the most impressive of the four Southern Birds. It lies south of Fomalhaut; one way to find it is to prolong a line from Beta and Alpha Pegasi, in the Square of Pegasus, through Fomalhaut. The leading stars are Alpha or Alnair (1.7), Beta or Al Dhanab (2.1) and Gamma (3.0). The general pattern is very distinctive, with a long line of stars extending from Gamma through Beta and the fainter stars Epsilon (3.5) and Zeta (4.1). Delta and Mu look like wide pairs, though in each case the components are not physically associated, and are too far apart to be classed as true doubles.

Alnair and Al Dhanab provide a good case of colour contrast. Alnair is a bluish-white B-type star, while Al Dhanab has an M-type spectrum and is a lovely orange. Iota (3.9) and Lambda (4.5) are also orange, as is well shown with × 7. Pi, somewhat away from the line joining Alnair to Al Dhanab, is an S-type red irregular variable with a range of from 5.8 to 6.4. There are several Mira variables in Grus within binocular range when near maximum (R, S and T), but for most of the time they are much too faint.

HERCULES: *Hercules*

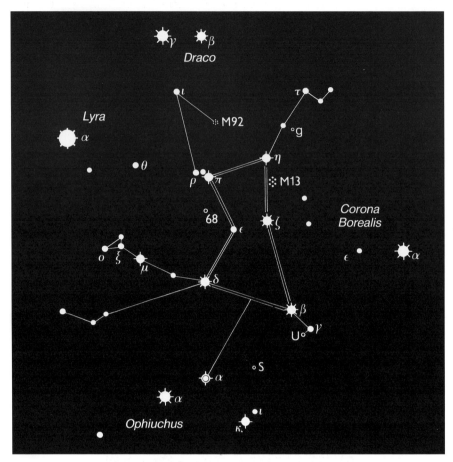

Hercules is a large constellation, but not a very bright one. It occupies the triangle bounded by Vega, Rasalhague (Alpha Ophiuchi) and Alphekka (Alpha Coronæ). Alpha Herculis, or Rasalgethi, lies close to Alpha Ophiuchi, rather divorced from the rest of the constellation. It is a huge red supergiant, 218 light-years away and over 700 times as luminous as the Sun. Like many M-type supergiants it is variable; the official range is from 3.0 to 3.8, but for most of the time it is comparable with Kappa Ophiuchi (3.2) which makes a convenient comparison star. There is said to be a rough period of about 100 days, but I have never been able to confirm this. The colour is very evident in binoculars The other leading stars of Hercules are Beta and Zeta (each 2.8), Delta (3.1), Pi (3.2), Mu (3.4) and Eta (3.5). Theta (3.9) has a K-type spectrum, and in binoculars appears to me as slightly 'off-white'.

U Herculis is a Mira variable near Gamma (3.7). It can attain magnitude 7, but it is not a suitable binocular object, as at minimum it falls to below 13.

The main objects of interest in Hercules are the two globular clusters, M13 and M92. M13 is the brighter, and is in fact the most prominent globular ever visible from latitudes such as that of Britain and the northern United States.

It is dimly discernible with the naked eye between Eta and Zeta, considerably closer to Eta. It was discovered by Edmond Halley in 1714; it is some 22 000 light-years away, with a real diameter of around 160 light-years. It is very evident in any binoculars; with × 20 I always feel that I can begin to resolve its outer edges. Oddly enough, it is rather poor in short-period variable stars.

M92, found by J.E. Bode in 1777, is fainter, on the fringe of naked-eye visibility; it is 35 000 light-years away, and about 90 light-years in diameter. With × 8 it is in the field with Iota and the Pi–Rho pair, and is obvious enough; with × 12 it is not so very much inferior to M13. With × 20 it is almost resolvable at its edges, but not quite – at least to my eyes.

On the whole Hercules is rather a barren constellation, but it is at least graced by the presence of these two splendid globulars.

M13, globular cluster in Hercules (Ben Mayer © 1980).

HOROLOGIUM: *the Clock*

A very dim constellation. Its brightest star, the orange K-type Alpha, is of magnitude 3.9; it is rather isolated, and forms a very wide pair with Delta (4.9). It is probably best found by looking directly between Achernar to the one side and Lepus to the other. The only object of any binocular interest is the long-period variable R Horologii, which may reach magnitude 4.7 at maximum although it drops to below 14 at minimum; the period is 403 days, and the spectrum is of type M, so that the star is very red It is not easy to locate, but Chi and Phi Eridani point to it. Horologium is far too southerly to rise over Britain or most of the United States. I have included it in the map with Eridanus.

HYDRA: *the Watersnake*

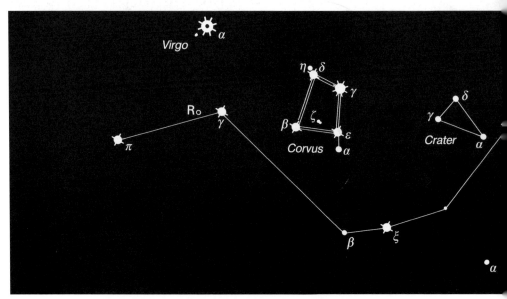

Now that Argo Navis has been dismantled, Hydra is the largest constellation in the sky, but it has only one bright star, the reddish Alpha or Alphard, nicknamed the 'Solitary One' because of its isolation; it has been suspected of variability, though officially the magnitude is given as constant at 2.0. Next come Gamma (3.0), Zeta and Nu (each 3.1), Pi (3.3), Epsilon (3.4) and Xi (3.5). Hydra sprawls from the boundary of Canis Minor through to the south of Corvus and Virgo. Alphard is easy to locate, because Castor and Pollux, in the Twins, point straight to it. Between Alphard and the Twins is the Watersnake's Head, made up of Zeta, Epsilon, Delta (4.2) and Eta (4.3). Zeta has a K-type spectrum, but even with × 20 I can see little colour in it. All the stars of the Head are in the same × 7 field.

R Hydræ, in the field with Gamma, is a Mira variable; as it can rise to the fourth magnitude it can be a binocular object, and is very red, though at minimum it falls to magnitude 11.

M48 is a bright galactic cluster. As the magnitude is above 6, it is detectable with the naked eye under very good conditions, but it is rather isolated, and takes time to identify. The 'guide' is the trio of stars c, 1 and 2 Hydræ, which are obvious in binoculars. Unfortunately they are just out of the × 7 field with Hydra's Head, but sweeping from the Head stars is a good way to find them. M48 lies in the × 7 field with the trio. It is clear enough, though large and dim; any higher magnification shows it well, though even with × 20 I am not at all sure that I can begin to resolve it into stars. Another method of locating it is to use Beta Canis Minoris and Procyon, which point to Zeta Monocerotis (4.3). Zeta, the trio, and M48 form a triangle, and are in the same × 7 field.

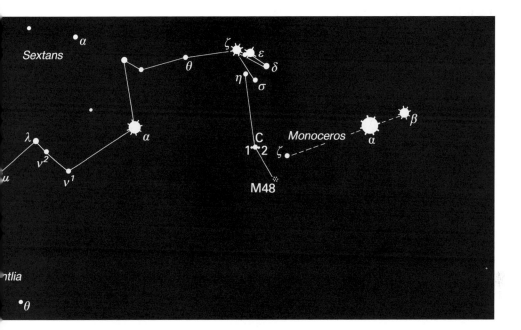

HYDRUS: *the Little Snake*

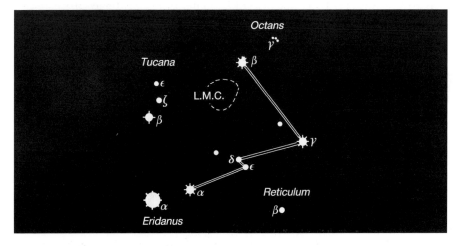

A far-southern constellation; the brightest stars are Beta (2.8), Alpha (2.9) and Gamma (3.2). Alpha is in the same × 7 field with Achernar.

Beta Hydri is the nearest fairly prominent star to the South Pole, but it is still over 12 degrees away from the polar point. With × 7, it is just in the same field with the little trio of Gamma Octantis, which is a good way of locating Octans. Hydrus itself is devoid of interesting objects.

It is worth remembering that in conditions of mist or moonlight, the whole south-polar region appears blank, and Beta Hydri is often the only star fairly close to the pole to remain visible with the naked eye.

INDUS: *the Indian*

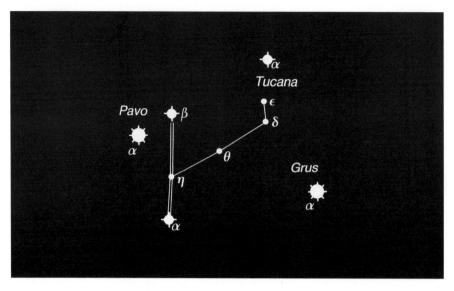

An undistinguished southern constellation. Its brightest star, Alpha (3.1) forms a triangle with Alpha Gruis and Alpha Pavonis; the only other star above the fourth magnitude, Beta (3.6), is close to Alpha Pavonis.

The only object of any interest is Epsilon (4.7), which, at a distance of 11.4 light-years, is one of the closest of the naked-eye stars. It is relatively small and cool, with only 13 per cent of the Sun's luminosity and a diameter of about 1 000 000 kilometres. It is the least luminous star visible with the naked eye. It is in the same × 7 field with Delta (4.4).

LACERTA: *the Lizard*

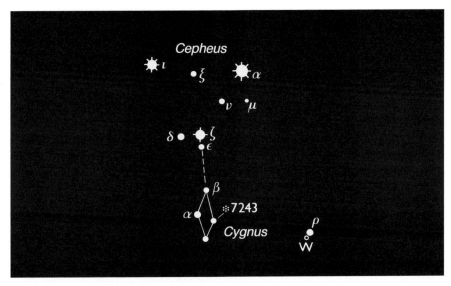

A faint northern constellation, adjoining Cepheus. There is only one star, Alpha (3.8) above the fourth magnitude. There is a small diamond of dim stars, of which Alpha is the brightest; to find them, use Zeta and Epsilon Cephei as guides. Epsilon Cephei is just in the × 7 field with Beta Lacertæ (4.4). Beta Lacertæ is a red star of type K.

The open cluster NGC 7243 (C16) forms an equilateral triangle with Alpha and Beta. Using × 20 I have suspected it – the integrated magnitude is given between 7 and 8 – but not with certainty, and there is nothing else of any real interest in Lacerta. A bright nova flared up here in 1936, but has long since become much too faint to be seen except with very large telescopes.

LEO: *the Lion*

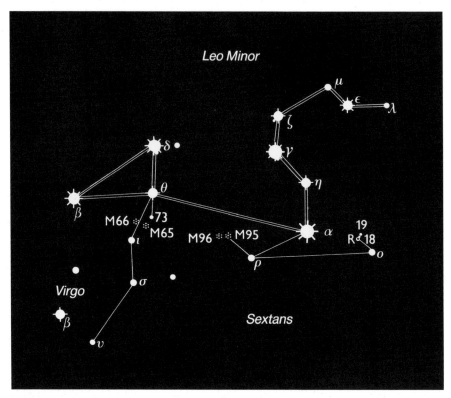

Leo is one of the brightest of the Zodiacal constellations. The leading stars are Alpha or Regulus (1.3), Gamma (2.1), Beta or Denebola (2.1), Delta (2.6), Epsilon (3.0), Theta (3.3), Zeta (3.4) and Eta and Omicron (each 3.5). The main pattern is the so-called Sickle, a curved arrangement of stars extending away from Regulus: Eta, Gamma, Zeta, Mu (3 9), Epsilon, and Lambda (4.3). The rest of Leo consists mainly of the triangle formed by Beta, Delta and Theta. Beta, often known by its old proper name, was ranked of the first magnitude in ancient times, but is at present slightly below the second. Since it is an A-type star there has probably been no real change, though it may be very slightly variable.

Gamma is an orange star of type K; telescopically it is a fine binary, but it cannot be split with binoculars. Two of the Sickle stars, Mu and Lambda, are also orange, and their colours are easy with × 7.

Near Regulus, binoculars show two inconspicuous stars, 18 Leonis (5.8) and 19 Leonis (6.4). Forming a group with them is the Mira variable R Leonis, which has a period of 312 days. It is characteristically red, with an M-type spectrum, and at maximum it may rise to magnitude 5.4, so that it is then an easy binocular object; with × 12 or higher magnification I have seen strong colour in it. At minimum it falls to magnitude 10.5. There are many galaxies in Leo. Most of them are faint, but two, M65 and M66, are within binocular range. They are in the same field with Theta; both are spirals, though of course their forms are

102

not visible with binoculars or small telescopes. It has been said that they show up as a striking pair with powers of × 12 or more. Frankly I always find them difficult even with × 20. The separation between them is 21 minutes of arc, and they are genuinely associated; they are around 29 000 000 light-years away from us, and 180 000 light-years apart. The guide star is 73 Leonis (5.5).

Two more galaxies, M95 and M96, lie east of Regulus, in the same × 7 field as Rho (3.8). These also are spirals. There are many reports of their having been seen with binoculars, but I have never had any success. Others may be more fortunate; at any rate, the two galaxies are worth seeking out.

LEO MINOR: *the Little Lion*

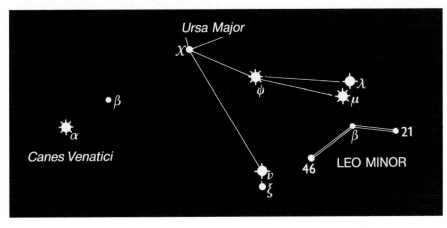

A small constellation with dubious claims to a separate identity. Its leading stars are 46 (3.8) and Beta (4.2).

Leo Minor lies between the Lambda–Mu and Xi–Nu pairs in Ursa Major. 46 Leonis Minoris is just in the same × 7 field with Xi and Nu, and is identifiable because it has a K-type spectrum and shows its orange colour when viewed through binoculars; unusually for so dim a star it has a proper name Præcipua. The Mira variable R Leonis Minoris can reach magnitude 6.3 at maximum and is then detectable with binoculars, but at minimum it sinks to below magnitude 12. Its period is 372 days. It is not too easy to find, but lies in more or less of a line with Beta and 21 (4.5). I have included Leo Minor in the map with Ursa Major.

LEPUS: *the Hare*

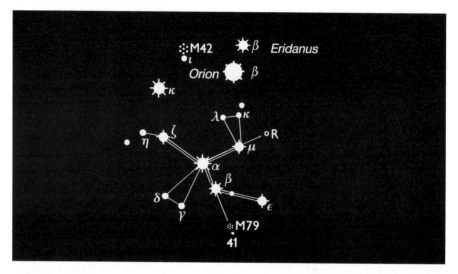

Lepus is a small but distinctive constellation, adjoining Orion to the south. Its leading stars are Alpha (2.6), Beta (2.8), Epsilon (3.2), Mu (3.3) and Zeta (3.5). Mu forms a triangle with two fainter stars, Lambda (4.3) and Kappa (4.4). In the same × 7 field is R Leporis, of spectral type N, a Mira variable with a period of 432 days. It never drops below magnitude 10.5, and at maximum it may rise to 5.9, so that it is then on the fringe of naked-eye visibility and is easy in any binoculars; it is distinguished by its intense redness, and it has been nicknamed the Crimson Star. With × 20 I have found that near maximum its colour is very clear, and it is detectable with lower magnifications. It is well over 1000 light-years away, and it is at least 500 times as luminous as the Sun, but its surface is very cool for a normal star – hence the crimson colour.

The other interesting binocular object in Lepus is the globular cluster M79. It is large, rich and compressed, but it is only of the eighth magnitude, and I find it very difficult with × 7 and not really easy with × 8.5 or × 12, though with × 20 it is obvious enough. To find it, follow a line from Alpha through Beta and extend it until you come to the next reasonably bright star, 41 Leporis (5.5). M79 is in the same field with 41, and against a dark sky there should be no real problem in locating it, though under moonlight conditions I cannot see it without a telescope. From British and northern United States latitudes it is always low down, so southern-hemisphere observers will find it a much easier object.

LIBRA: *the Balance*

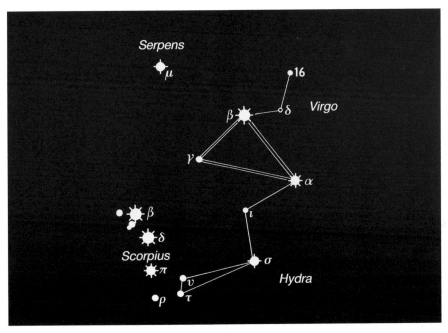

Libra is one of the least striking of the Zodiacal constellations. It adjoins Scorpius, and the star now known as Sigma Libræ was formerly included in that constellation, as Gamma Scorpii. The chief stars of Libra are Beta or Zubenelchemale (2.6), Alpha or Zubenelgenubi (2.7) and Sigma or Zubenalgenubi (3.3). The old name for Libra itself was Chelæ Scorpionis, the Scorpion's Claws, and the strange names of the leading stars indicate those claws.

Frankly, there is nothing of much interest in the constellation. Delta Libræ is an Algol-type eclipsing binary, with a range of from 4.8 to 6.1 and a period of 2.33 days. It makes a triangle with Beta and 16 Libræ (4.5); using × 7, Delta and Beta are in the same field, but Beta and 16 are not. Beta has a B8-type spectrum, and is about 100 times as luminous as the Sun; it is often said to be the only naked-eye star which is greenish in colour, though I have never noted this either with or without optical aid, and to me Beta always looks white. Alpha is made up of two components of magnitudes 3 and 5; the separation is 231 seconds of arc, so that binoculars give a good view of the pair even though it is not in the least prominent.

105

LUPUS: *the Wolf*

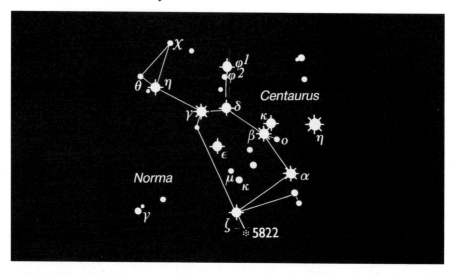

I always find this a confusing constellation. It adjoins Centaurus, and contains a number of brightish stars: Alpha (2.3), Beta (2.7), Gamma (2.8) and Delta (3.2), plus several more above the fourth magnitude. There is no definite pattern, and very little of interest. However the open cluster NGC 5822, with an integrated magnitude of above 7, is easy to find; it lies in the field with Zeta. It may be worth looking at the pair made up of Phi[1] (3.6) which is orange, and Phi[2] (4.5) which is white.

LYNX: *the Lynx*

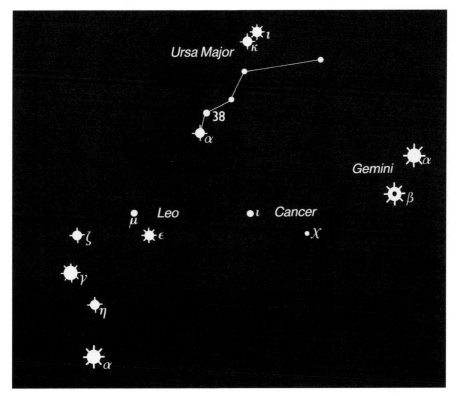

A very obscure constellation, lying mainly between Ursa Major and Gemini. Its two brightest stars, Alpha (3.1) and 38 (3.9) are in the same × 7 field. Alpha is not hard to find, because it is decidedly isolated and forms an equilateral triangle with Regulus and Pollux; it is of type M, and very red as seen with any binoculars. Lynx contains nothing else of any note.

LYRA: *the Lyre*

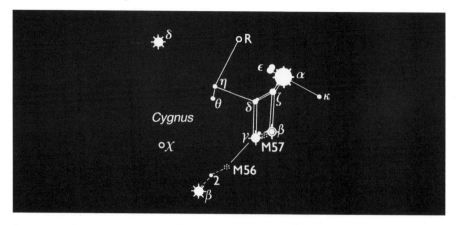

Though Lyra is small, it contains a number of interesting objects, and of course it is graced by the presence of Vega, the lovely blue star which is almost overhead as seen from British latitudes during summer evenings. It is 26 light-years away and 52 times as luminous as the Sun; it is certainly the bluest of the brilliant stars. Its magnitude is 0.0 so that it is surpassed only by Sirius, Canopus, Alpha Centauri and Arcturus. In 1983, observations from the IRAS infra-red satellite showed that it is associated with cool material which might be interpreted as a planetary system, though it would be most unwise to jump to any conclusions. At any rate, Vega is a glorious sight in any pair of binoculars.

The little quadrilateral near Vega is made up of Gamma (3.2), Zeta and Delta (each 4.3) and the eclipsing binary Beta. which has a range of from 3.4 to 4.3 and exhibits alternate deep and shallow minima. The full period is 12.9 days. To the astrophysicist it is one of the most interesting objects in the sky, though the naked-eye or binocular observer must be content with watching its fluctuations; Gamma and Zeta are suitable comparison stars. The separation between the two components is only about 35 000 000 kilometres, so that no telescope can show them separately, and each must be egg-shaped.

The other variable suited to binocular observation is the red semi-regular R Lyræ, whose colour is evident even with a low magnification. The range is from 4.0 to 5.0, and there is a rough period of around 47 days. It is an awkward object to observe, because the only useful comparison stars, the white Eta and the slightly orange Theta, are just outside its field with × 7 binoculars, and there is nothing else available. Incidentally, both Eta and Theta are officially listed as being of magnitude 4.4, but I find that Theta is the brighter of the two by about 0.2 of a magnitude. Since Theta has a K-type spectrum, I suppose that it may be slightly variable.

Epsilon Lyræ, close to Vega, is the famous 'double-double' star. The components are of magnitudes 4.7 and 5.1, and are 208 seconds of arc apart, so that any binoculars will split them, and keen-eyed people can see both with no optical aid at all. Telescopically, each component is again double. The system is a genuine one, and is decidedly complicated. The true separation between the bright pairs is about 0.2 of a light-year, and the distance from Earth is 180 light-years.

Zeta is also double. The components are of magnitudes 4.4 and 5.7, and since the separation is almost 44 seconds of arc it should be possible to see them individually with × 20 binoculars, but I have never been able to do so with certainty.

Delta is another double, with a separation of over 10 minutes of arc. The brighter of the two (4.5) is of type M, and its obvious redness makes a good contrast with its companion (5.5), which is white. Even × 7 shows it well, and with higher magnifications the pair is striking, particularly as the whole adjacent field is rich.

Directly between Beta and Gamma lies the famous Ring Nebula (M57), one of the brightest of the planetaries. It has been claimed that binoculars will show it, but again I must admit failure. I can however see the globular cluster M56; it is difficult with × 7, detectable without much trouble with × 12, and easy with × 20. To locate it, begin at Beta Cygni and then identify 2 Cygni (4.9) which lies in the direction of Vega. Using × 20, M56 is in the same field as 2, and with × 12 the cluster, 2, and Beta Cygni are all included. M56 is not at all prominent, but shows up as a faint patch of light. The official integrated magnitude is 8, but to me it seems slightly brighter than this, and giving definite magnitude values for diffuse objects is always a problem.

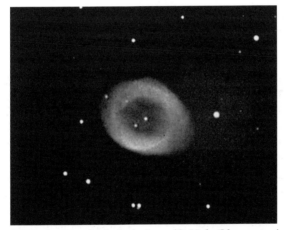

M57, the Ring Nebula in Lyra (© Hale Observatories).

MENSA: *the Table*

Originally Mons Mensæ, the Table Mountain. It has the dubious distinction of being the only constellation with no star as bright as the fifth magnitude; its main star. Alpha (5.2) lies between Volans and Gamma Hydri.

A small part of the Large Cloud of Magellan extends into Mensa, but otherwise there is absolutely nothing here of interest. I have included it in the map with Dorado.

MICROSCOPIUM: *the Microscope*

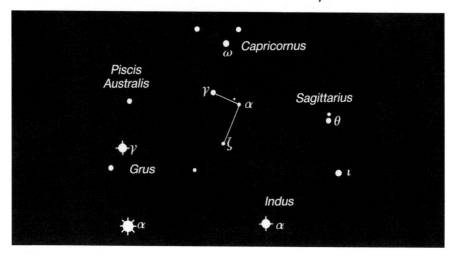

A totally unremarkable constellation south of Capricornus, more or less between Capricornus and Grus. Its brightest star, Gamma, is only of magnitude 4.7, and there are no objects of any interest at all.

MONOCEROS: *the Unicorn*

The celestial Unicorn is a fairly large but faint constellation, filling most of the large triangle outlined by Betelgeux, Sirius and Procyon; it is crossed by the equator. The brightest star, Beta is only of magnitude 3.7, and there is no distinctive shape, but the Milky Way runs right through it, so that there are plenty of rich fields. Gamma (4.0) is somewhat orange.

Probably the most interesting object is the open cluster NGC 2244 (C50) round the star 12 Monocerotis (5.8); it is quite easy to locate with binoculars, and I am rather surprised that it is not included in the Messier list. Find Epsilon Monocerotis (4.3), which lies slightly off the line joining Procyon to Betelgeux, considerably closer to Betelgeux. The cluster lies in the same × 7 or × 8.5 field, together with a fainter star, 13 Monocerotis (4.5). The cluster is made up of a small quadrilateral which is distinctive enough, and is very characteristic with × 20.

Surrounding the cluster is the famous Rosette Nebula, NGC 2237 (C49). Photographed with large telescopes, this is one of the loveliest and most vividly-coloured of all the nebulæ, but visually it is not at all easy to see. I am told that with good binoculars (presumably × 12 or more) it can be made out as an aura of soft light surrounding the cluster, but I have had no success myself.

The open cluster M50 is less spectacular, but easy to locate. Start from Sirius, and move on to Theta Canis Majoris (4.1) which is noticeably orange, and is in the Sirius field with × 8.5, though just out of it with × 20. Continue along the same line for an equal distance, and you will come to M50. It appears as a dim blur with low magnification, but with × 20 I can see individual stars in it. Its real diameter is about 13 light-years.

Beyond M50 we come to Alpha Monocerotis (3.9). In the same field lies the open cluster NGC 2506 (C54), easy with × 7 and large with × 20, though with few individual stars. It is very close to the border between Monoceros and Puppis, and there is nothing particularly noteworthy about it.

NGC 2237, the Rosette Nebula in Monoceros (© Royal Observatory Edinburgh).

MUSCA AUSTRALIS: *the Southern Fly*

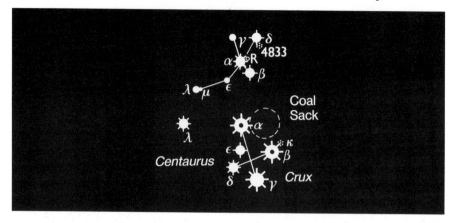

This small but fairly distinctive constellation, adjoining the Southern Cross, is now known generally as Musca without the 'Australis'. The brightest stars are Alpha (2.7), Beta (3.0), and Delta and Lambda (each 3.6). Alpha, Beta and Delta are in the same × 12 field.

Lambda and Mu (4.7) are just in the × 12 field with Alpha Crucis. They make up a wide pair, with beautiful contrasting colours; Lambda is white and Mu very red, with an M-type spectrum. The field is very striking with × 7. Epsilon (4.1), between Beta and Lambda, is also of type M, and is decidedly orange, though to me the colour is much less pronounced than that of Mu.

The globular cluster NGC 4833 (C105) lies close to Delta. It is said to be visible with binoculars, but I have never been able to see it without a telescope.

NORMA: *the Rule*

Formerly Quadrans Euclidis, Euclid's Quadrant. Norma is very obscure; its brightest star, Gamma, is only of magnitude 4.0. The only subject of any interest is the open cluster NGC 6087 (C89), which lies between Alpha Centauri and Zeta Aræ, closer to the latter. It is fairly rich, and is an easy binocular object; it contains the Cepheid variable S Normæ, which has a range of from 6.8 to 7.8 and a period of 9.7 days. I have included Norma in the map with Ara.

OCTANS: *the Octant*

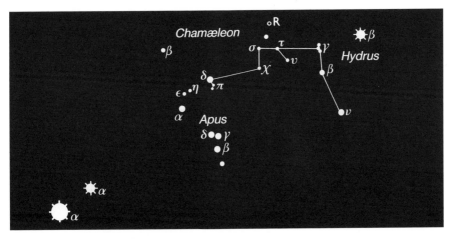

Octans contains the South Pole of the sky. Unfortunately it is a very barren group, and when I first went to the southern hemisphere I had difficulty in identifying the pole star, Sigma Octantis, which is only of magnitude 5.5, and is none too easy to see with the naked eye unless the sky is really dark and clear. The only star in Octans above the fourth magnitude is Nu (3.8), which is orange and of type K.

As we have noted, the South Pole lies about midway between Achernar and the Southern Cross but this is only a rough guide, and the nearest bright-ish star to the polar point, Beta Hydri, is some distance away. So after a little 'finding my way around', I hit upon what I think is the best way to find the pole, using × 7 binoculars.

First identify Alpha Apodis, which lies in the direction of a line through Alpha Centauri and Alpha Circini. In the same field as Alpha Apodis are the two faint stars Epsilon Apodis (5.2) and Eta Apodis (5.0). These point straight to Delta Octantis (4.3), which is clearly orange, and has two dim stars, Pi[1] and Pi[2] Octantis, close beside it. Now put Delta Octantis in the edge of the field, and continue the line from Apus. Chi Octantis (3.2) will be on the far side of the field; centre it, and then you will see two more stars of about the same brightness, Sigma and Tau. These three are in the same field, and make up a triangle. The south polar star, Sigma, is the second in order from Delta. Using × 12, the three are in the same field together with Upsilon (5.7).

This may sound complicated, but once Sigma has been identified it is easy to find again with binoculars, though much less so with the naked eye. It is almost one degree away from the polar point; it is 120 light-years away and only about six times as luminous as the Sun, so that it is by no means the equal of the northern hemisphere's Polaris.

The little trio of Gamma[1], Gamma[2] and Gamma[3] Octantis (5.1, 5.7 and 5.3, respectively) is in the same × 7 field with Epsilon, and is also just in the field with Beta Hydri, which is an equally good way of locating Octans. A magnification of × 7 is just wide enough to include Gamma and Nu. The Mira variable R rises to 6.4 at maximum.

OPHIUCHUS: *the Serpent-bearer*

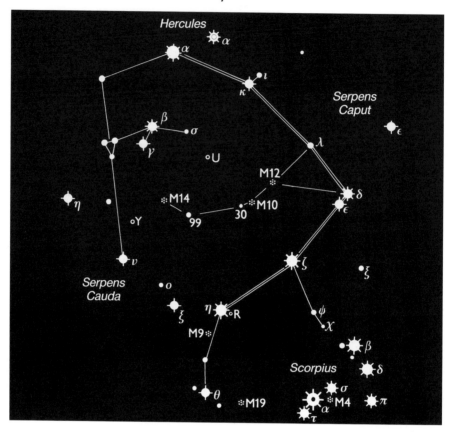

A large constellation, which straddles the equator and actually intrudes into the Zodiac between Scorpius and Sagittarius. The leading stars are Alpha or Rasalhague (2.1), Eta (2.4), Zeta (2.6), Delta (2.7), Beta (2.8), Kappa and Epsilon (each 3.2) and Theta (3.3). Beta is orange, with a K-type spectrum; so is Kappa, which makes a good comparison for the red variable Alpha Herculis. However, the reddest of the leaders of Ophiuchus is Delta, which is type M and whose colour is striking in binoculars; it forms a pair with Epsilon.

Chi Ophiuchi is an irregular variable with a range of from 4 to 5; it lies close to Phi (4.3) which makes a good comparison star. It is only just inside Ophiuchus, near the Scorpion's head. R Ophiuchi, a Mira star, lies in the field with Eta; it can rise to the seventh magnitude, but for most of its 302-day period it is far below binocular range.

There are various globular clusters in Ophiuchus. M10 and M12 form a pair. I have great difficulty in picking them up with binoculars, but with × 7 I find that the best method is to start at the Delta–Epsilon pair and then proceed to Lambda (3.8); M12 forms a triangle with Lambda and Epsilon. M10 is nearby, close to the reddish star 30 Ophiuchi (5.0). The main problem is that neither cluster is at all prominent, and it is easy to be mislead by the adjacent star-fields. Other globulars are M19, M9 and M14.

ORION: *the Hunter*

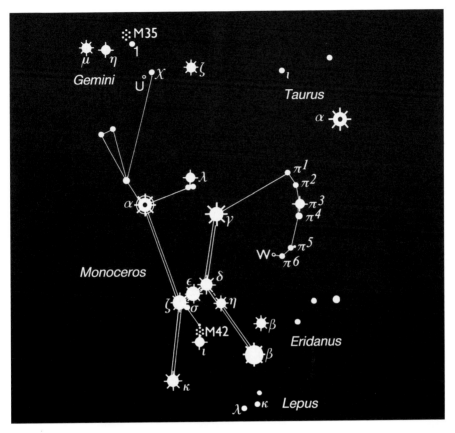

This is arguably the most splendid of all the constellations. Its leading stars are Beta or Rigel (0.1), Alpha or Betelgeux (variable between 0.1 and 0.9), Gamma or Bellatrix (1.6), Epsilon or Alnilam (1.7), Zeta or Alnitak (1.8), Kappa or Saiph (2.1), Delta or Mintaka (2.3, very slightly variable), Iota (2.8), Pi^3 (3.2), and Eta and Lambda (each 3.4).

Orion is a particularly useful direction-finder. It has the added advantage that the celestial equator passes through it, narrowly missing Mintaka in the Hunter's Belt, so that it is visible from every inhabited part of the world. The problem in a book of this kind is, of course, that in the northern hemisphere Betelgeux is higher than Rigel, while from the southern hemisphere the reverse is true, and from the equator the Hunter appears in a rather undignified sideways position. Since I happen to be English, I have oriented the chart here for northern observers, but I hope that those in the south will forgive me.

The two leading stars, Betelgeux and Rigel, are very different. Betelgeux is a vast red supergiant, large enough to swallow up the whole orbit of the Earth round the Sun, while the pure white Rigel is a cosmic searchlight with 60 000 times the Sun's luminosity. With the naked eye, the colour difference is very evident, and binoculars bring it out splendidly. Betelgeux is decidedly variable. It swells and shrinks, changing its output as it does so; the

ORION

official magnitude range is from 0.4 to 0.9, but over the past decades I have seen it matching Rigel on rare occasions, while it is not often as faint as Aldebaran. Estimates are not too easy, because the only useful comparison stars – Procyon, Aldebaran and Rigel – are always at different altitudes, and allowance has to be made for extinction (that is to say, the dimming of a low-altitude star due to the Earth's atmosphere). There is a rough period which is said to be around 2070 days or just over 5½ years, but it *is* rough. Rigel is only marginally fainter than Capella or Vega, and is much more luminous than either. Its distance is of the order of 900 light-years, nearly twice that of Betelgeux.

All the other stars of the main pattern are hot, white and powerful; indeed Saiph, over 2000 light-years away, is almost as luminous as Rigel. (Calculations show that a million years ago it shone as much the brightest star in the sky, and equalled Venus; it was closer to us then than it is now.) Leading away from the Belt is the Hunter's Sword, containing the Great Nebula, M42. With the naked eye M42 is visible as a misty patch, and binoculars show it excellently. With $\times$ 7 or higher magnification it is possible to see the stars mixed in with it, and $\times$ 20 reveals the Trapezium – the celebrated multiple star Theta Orionis, which lies just on our side of the Nebula and is responsible for making it shine. It is also easy to see the patches of dark nebulosity which are not illuminated by any convenient stars and make their presence known by blotting out the light of objects beyond.

With $\times$ 7 or $\times$ 8.5, M42 is in the same field as the Belt, making up a spectacle which to me seems unrivalled in the whole of the sky. The Nebula is over 1000 light-years away, and we know that fresh stars are being formed inside it, because one at least has been 'caught in the act' as it started to emit visible light. In fact M42 is only the brightest part of a vast nebular cloud which covers most of Orion, but I have never been able to see the rest of it with binoculars. M42 itself is comparatively young, probably not more than 30 000 years old. It was not recorded before the year 1610, and there have been suggestions that it brightened up abruptly at that time, though personally I am sceptical; it would be a strange coincidence if the Nebula burst into prominence at just the time when mankind invented the telescope!

The red semi-regular variable W Orionis is in the same field with Pi^6 (4.5) – the southernmost member of a line of stars all of which, for some illogical reason, are lettered Pi. It has an N-type spectrum, and a range of from 5.9 to 7.7, with a rough period of 212 days. It is easy to identify, because its obvious colour makes it stand out at once. Actually it is redder than Betelgeux, but it looks so much fainter that the colour is not so striking even in binoculars. U Orionis, on the borders of Orion and Taurus, is a Mira star with a range of from 5.3 to 12.6, so that near maximum it is an easy binocular object; it is a member of a well-marked little group lying between Zeta Tauri and Eta Geminorum. Like all Mira stars it is red, and with binoculars the colour is detectable when the star is near maximum.

The whole of Orion is rich, and is worth sweeping. Using $\times$ 7, the region of Alnitak is particularly rewarding.

PAVO: *the Peacock*

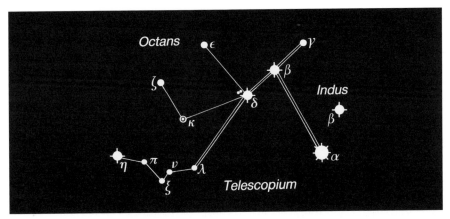

Pavo, the second of the Southern Birds, is not so distinctive as Grus, but it is not hard to find, because Alpha Centauri and Alpha Trianguli Australe show the way to its brightest star, Alpha, which is bluish-white and is of magnitude 1.9. Actually Alpha is rather isolated from the rest of the constellation. The other leading stars are Beta (3.4), Delta and Eta (each 3.6), Kappa (variable) and Epsilon and Zeta (each 4.0). Xi (4.4) is an M-type red star.

The most notable object in Pavo is the short-period variable Kappa, which is a Type II Cepheid, rather less luminous than an ordinary Cepheid of the same period. The range is from 3.9 to 4.8, and the period is 9.1 days; suitable comparison stars are Zeta, Pi (4.4) and Nu (4.6). Lambda is also variable, but of uncertain type. It has a B-type spectrum, and is usually just below the fourth magnitude.

PEGASUS: *the Flying Horse*

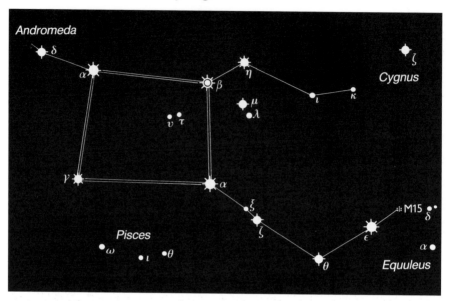

Pegasus is the main autumn constellation as seen from northern latitudes. It is unmistakable, though in case of doubt two of the stars in the W of Cassiopeia can be used as guides. Of the stars in the Square, the brightest is Alpheratz (2.1), which used to be known as Delta Pegasi, but has now been transferred to Andromeda as Alpha Andromodæ.

In Pegasus itself, the main stars are Epsilon (2.4), Beta (variable), Alpha (2.5), Gamma (2.8), Eta (2.9) and Zeta (3.4). Epsilon, or Enif, is well away from the Square, close to Equuleus. It is reddish, with a K-type spectrum, and has been strongly suspected of variability; it is over 500 light-years away, and 5000 times as luminous as the Sun. It is always worth checking its brightness against Alpha Pegasi (Markab), which is white and definitely constant.

Beta Pegasi, or Scheat, is an M-type semi-regular variable, with a rough period of around 36 days. Its orange colour is quite evident with the naked eye, and striking in binoculars. Markab and Gamma (Algenib) make useful comparisons, though one has to allow for differences in altitude. The range is from about 2.4 to 2.9.

The most interesting binocular object is M15, a globular cluster not far below naked-eye visibility; it was discovered by Maraldi in 1746, while he was looking for a comet. It has a condensed centre, and in binoculars it looks like a circular fuzz. It lies in a direct line with Theta (3.7) and Epsilon, and there is a sixth-magnitude star close beside it. It is easy to identify with × 7, and with × 20 it is not greatly inferior to M13 in Hercules.

Recently I went outside on a clear night, and started to count the stars I could see inside the Square; using × 7. I gave up! The two brightest stars in the Square are Tau (4.6) and Upsilon (4.4).

PERSEUS: *Perseus*

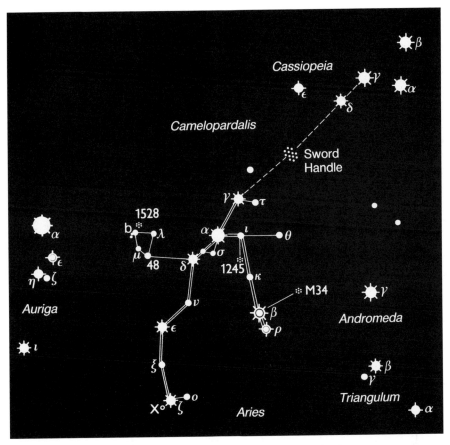

A large, impressive constellation between Auriga and Andromeda. It is crossed by the Milky Way, and contains many glorious binocular fields. The leading stars are Alpha or Mirphak (1.8), Beta or Algol (2.1 at maximum), Zeta (2.8), Epsilon and Gamma (each 2.9), Delta (3.0) and Rho (variable, about 3.2 at maximum). Mirphak is an F5-type giant, 620 light-years away and 6000 times as luminous as the Sun. It is said to be slightly yellowish, though I have never seen any colour in it either with the naked eye or with binoculars.

Algol, the prototype eclipsing binary, has been described earlier. The range is from 2.1 to 3.3, and the period 2.7 days; the secondary minimum, when the fainter component is eclipsed by the brighter, is too slight to be detected except with sensitive equipment. There is no shortage of comparison stars, and the times of minima are given in almanacs. If you look at Algol and find that it is no brighter than Epsilon or Zeta, you may be sure that an eclipse is in progress.

To either side of Algol are Kappa (3.8) and Rho, which is an M-type semi-regular variable. The range is said to be from 3.2 to 4.2 and the period around 45 days, but I have never been able to make much periodicity out of it, and neither have I ever seen it brighter than magnitude 3.6.

119

PERSEUS

Zeta Persei is very luminous and remote (over 15 000 times as powerful as the Sun). In the same × 7 field are Omicron (3.8) and the variable X Persei, which seems to be irregular, and fluctuates between magnitudes 6.0 and 6.6. See the special chart. It is of special note because it is a source of X-rays, and is decidedly unusual. It is worth watching.

Perseus contains the superb Sword-Handle, NGC 869/864 (C14), otherwise known as Chi-h Persei. It consists of two clusters in the same binocular field. They are extremely easy to identify, between Mirphak and the W of Cassiopeia, and they are naked-eye objects on a clear night. Binoculars show them well. The distance is thought to be about 8000 light-years, and each cluster is 70 light-years in diameter.

M34 is an open cluster in the same × 12 field with Algol, forming a triangle with Algol and Rho. It is easy to see with × 7 as a hazy patch, and with × 12 there is a hint of resolution into stars; many individual stars can be seen in it with × 20. It is over 1000 light-years away.

Another open cluster, NGC 1245, is in the × 8.5 field with Mirphak and Kappa. It is by no means conspicuous, and is so scattered that identification can be a problem even with × 20. NGC 1528, yet another open cluster, is difficult with × 7, easy with × 12 and partly resolved with × 20. It lies within three degrees of Mirphak, and may be found by locating the little quadrilateral made up of 48, Mu, b and Lambda; the cluster lies just outside the quadrilateral, near b. It contains at least 40 stars, but binoculars show it only as a dim, slightly elliptical object of low surface brightness.

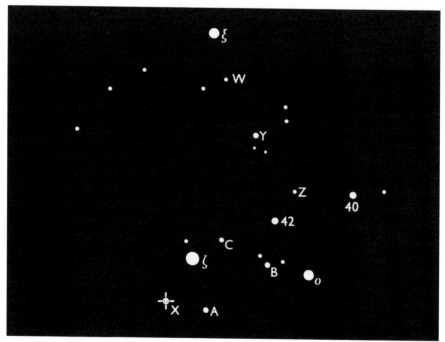

X Persei.

PHŒNIX: *the Phœnix*

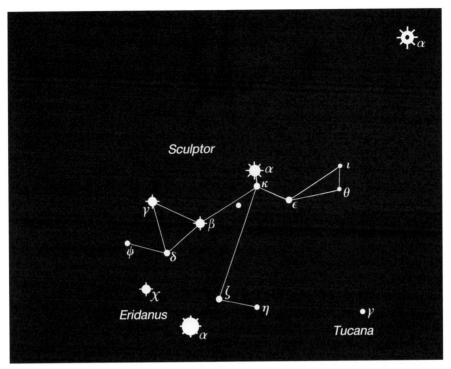

A Southern Bird, lying south of declination −40 degrees and therefore invisible from Britain and the northern United States. The leading stars are Alpha or Ankaa (2.4), Beta (3.3), Gamma (3.4), Zeta (variable), and Epsilon, Kappa and Delta (each 3.9). Ankaa lies slightly off the mid-point of a line joining Fomalhaut to Achernar, which is probably the best way of identifying it. It is in the same × 7 field with Kappa. Epsilon and Psi (4.4) are orange when seen in binoculars.

Zeta is an Algol-type eclipsing binary, with a range of from 3.6 to 4.1 and a period of 1.67 days; suitable comparisons are Eta (4.4), Delta and Chi Eridani (3.7). In the same field as Iota (4.7) lies the rapid variable SX Phœnicis, which has a period of only 79 minutes – one of the shortest known. However, it has a very small range (7.1 to 7.3) and is not of real interest to the binocular observer.

PICTOR: *the Painter*

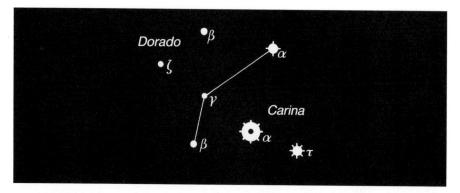

Originally Equuleus Pictoris (the Painter's Easel). It lies close to Canopus, and there are only two stars above the fourth magnitude: Alpha (3.3) and Beta (3.8). Beta, at a distance of 78 light-years, is of special note because it seems to be associated with material which could possibly indicate a planetary system, and has actually been photographed from the Las Campanas Observatory in Chile, though to the ordinary observer Beta looks like a normal white star.

A bright nova flared up in Pictor in 1925, and reached the first magnitude, but has long since become very faint. Otherwise there is nothing here of immediate interest.

PISCES: *the Fishes*

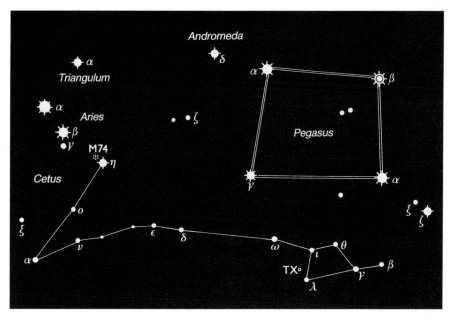

This is officially the last of the Zodiacal constellations, though since it now contains the Vernal Equinox it really ought to be the first. Apart from this it has little to recommend it, as it consists of a chain of dim stars extending from near the head of Cetus into the region south of the Square of Pegasus. Its brightest stars are Eta (3.6), Gamma (3.7), Alpha (3.8) and Omega (4.0).

The only object really worth seeking out with binoculars is the very red variable TX or 19 Piscium, which fluctuates irregularly between magnitudes 4.3 and 5.1; it has an N-type spectrum. To find it, first identify the little group made up of Gamma, Theta, Iota and Lambda, all of which are around the fourth magnitude. TX is in the same × 7 field with Iota and Lambda, forming a triangle with them. The colour is really pronounced, and with × 20 the star looks almost as vividly-hued as Mu Cephei.

Of the other stars in Pisces, Theta and Omicron (each 4.3) have K-type spectra; I can see that Omicron is slightly off-white, but I have never been able to see any colour at all in Theta.

Close to Eta is one of Messier's objects: M74, a spiral galaxy. I have heard reports that it can be glimpsed in binoculars, but even with my × 20 I have been unable to find it – though it is clear enough in a 15-cm reflecting telescope, and I can just make it out with my 7.6-cm refractor.

PISCIS AUSTRALIS: *the Southern Fish*

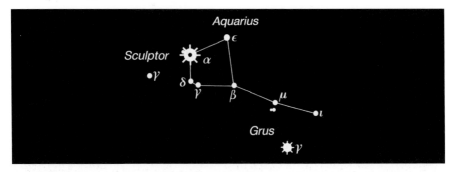

Otherwise known as Piscis Austrinus. It contains one first-magnitude star, Fomalhaut (1.2), and no others above 4.4. Fomalhaut is easy to find by using Scheat and Markab, in the Square of Pegasus, as pointers. (Beware of confusing Fomalhaut with Diphda or Beta Ceti, which is roughly aligned with the other two stars in the Square, Alpheratz and Algenib; but Beta Ceti is a magnitude the fainter of the two.)

Fomalhaut is a pure white star, 13 times as luminous as the Sun and 22 light-years away; in 1983 the IRAS satellite found that it is associated with cool material which may be planet-forming.

From Britain and the northern United States Fomalhaut is always very low down; but when near the zenith, as seen from countries south of the equator, it is very prominent – particularly because there are no other bright stars anywhere near it. The Southern Fish contains nothing else of note.

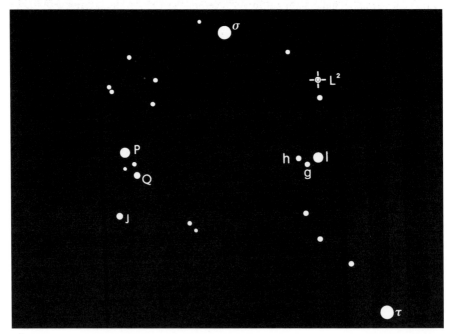

L^2 *Puppis.*

PUPPIS: *the Poop*

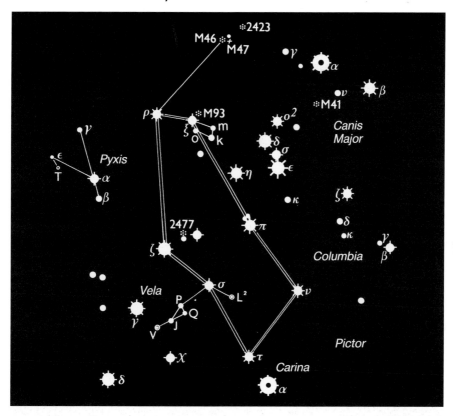

Puppis is one of the parts of the old Argo Navis. Its leading stars are Zeta (2.2), Pi (2.7), Rho (2.8), Tau (2.9), Nu and Sigma (each 3.2) and Xi (3.3).

Part of Puppis rises over Europe and the northern United States – notably Rho, which is not far from the prominent triangle made up of Eta, Delta and Epsilon Canis Majoris. Zeta, below the British horizon, has an O-type spectrum, and is highly luminous, with at least 55 000 times the output of the Sun.

L^2 Puppis is a semi-regular variable with a range of from 3.4 to 6.2, so that it is almost always visible with the naked eye and is well within binocular range even at minimum; it is of type M, and clearly orange. It is in the same field with Sigma. Not far off, and actually in the same × 7 field with Gamma Velorum, is V Puppis, an eclipsing binary of the Beta Lyræ type; the range is from 4.3 to 5.1, and the period is 1.45 days. Good comparison stars for it are J (3.6) and Q (4.7). The system is made up of two very luminous B-type giants almost in contact, so that no telescope will show them separately.

There are several bright open clusters in Puppis: M47, M46, M93, NGC 2423 and NGC 2571. M46 and M47 are close together, more or less in line with Beta Canis Majoris and Sirius; they are not spectacular, but are easy to locate. M93, in the × 7 field with Xi Puppis, is similar. NGC 2477 (C71) is virtually inaccessible from most of Europe, but it is easy to find from more southerly latitudes, as it is close to Zeta.

PYXIS: *the Mariner's Compass*

A small constellation, with only two stars as bright as the fourth magnitude: Alpha (3.7) and Beta (4.0). These two lie close together, and make up an equilateral triangle with Zeta Puppis and Lambda Velorum.

The only object of interest is the recurrent nova T Pyxidis. Normally it is very faint – below magnitude 14 – but it has shown outbursts in 1890, 1902, 1944 and 1966, when it has reached magnitude 7. Since one never knows when it may flare up again, there is good reason for watching the area. Find Epsilon Pyxidis (5.6) and memorize the × 7 field. If you see any star of comparable brightness, you may be sure that it is T.

Pyxis is included on the chart with Puppis.

RETICULUM: *the Net*

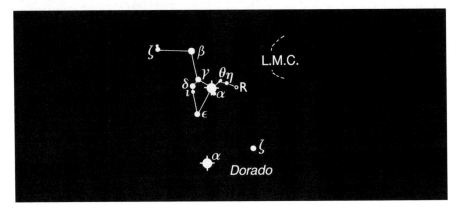

A far-southern constellation, not hard to find because it is so compact; it lies more or less between Achernar and Canopus. The brightest stars are Alpha (3.3) and Beta (3.8); Beta is orange, and so are Gamma (4.5), Epsilon (4.4) and Delta (4.6), so that the overall impression is quite distinctive. The Mira variable R Reticuli lies in the field with Alpha and also two fainter stars, Theta (6.2) and Eta (5.2). It can rise to magnitude 6.8, and can be distinguished by its redness, but for most of its 278-day period it is out of binocular range.

SAGITTA: *the Arrow*

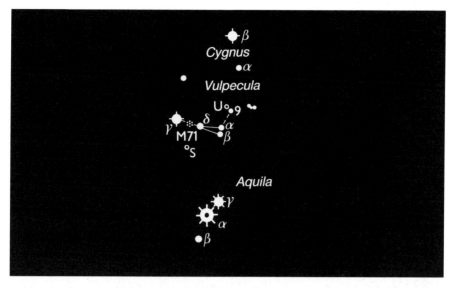

Sagitta is a very small constellation, but has a distinctive shape and is therefore easy to find; it lies directly between Altair and Beta Cygni. The brightest star, Gamma, (3.5), is of type K5, and binoculars show it as orange; the other leading stars, Delta (3.8) and Alpha and Beta (each 4.4) are white. The shape of the group really does give some impression of an arrow!

U Sagittæ is an Algol-type eclipsing binary. As its range is between 6.4 and 9.0, it is visible with binoculars when near maximum; the period is 3.4 days. To find it, use Beta and Alpha to show the way to 9 Sagittæ and then turn to a little cluster of stars which is actually in Vulpecula; U Sagittæ lies beyond. However, it will be of little real interest to the binocular observer.

The cluster M71 lies between Delta and Gamma. Its magnitude is only 9 according to official estimates; I have never seen it with low-power binoculars, but I have suspected it with × 20 though I would not have noticed it if I had not known just where it was. The NGC number is 6838. It is very distant, at 18 000 light-years, and there have been suggestions that it may be globular rather than a loose cluster.

SAGITTARIUS: *the Archer*

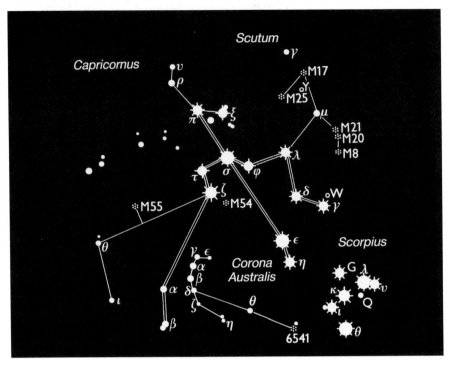

Sagittarius is the southernmost of the Zodiacal constellations, and is never well seen from the latitude of Britain or the northern United States; part of it never rises at all. From Australia or South Africa it is glorious, as it may pass overhead. The leading stars are Epsilon (1.8), Sigma (2.0), Zeta (2.6), Delta (2.7), Lambda (2.8), Pi and Gamma (each 3.0), Eta (3.1), Phi (3.2) and Tau (3.3). Curiously, the stars lettered Alpha and Beta are relatively faint (4.0 and 3.9 respectively). Gamma, Lambda, and Tau are orange; Lambda lies in a particularly rich field, best seen with × 7. Beta is made up of two components, separable with binoculars.

Sagittarius has no distinctive shape. Some people have likened it to a teapot, though I have never been able to make out why. It is, of course, celebrated for the magnificent star-clouds which indicate the direction of the centre of the Galaxy, and sweeping with binoculars over the whole region is very rewarding.

Observers in northern latitudes can never really appreciate how superb the star-clouds really are; when seen at high altitude they are striking by any standards. Beyond them lies that very mysterious region, the centre of the Galaxy, about which we know very little at the moment. Estimates of the distance of the galactic centre vary somewhat, but 30 000 light-years is not far from the truth. The Sun takes 225 000 000 years to complete one orbit round the centre – a period sometimes called the 'cosmic year'.

Clusters, both open and globular, abound. There are also two superb nebulæ M20 (the Trifid) and M8 (the Lagoon). M8, not far from Lambda, is an easy binocular object, though I always find its neighbour, M20, rather elusive. Photographs taken with large telescopes are needed to bring out their vivid colours, and with binoculars they appear white and milky. M17, the Omega Nebula, is of the same type; it lies at the edge of Sagittarius, and is in the same × 7 field with Gamma Scuti. The open cluster M21 is in the field with Mu (3.9), and another open cluster M25, forms a triangle with Mu Sagittarii and Gamma Scuti; the globular clusters M22 and M28 lie close to Lambda. M24, between Mu Sagittarii and Gamma Scuti, is not a true cluster, and is only a part of the Milky Way; M54 and M55 are globulars in the × 7 field with Zeta. The sparse open cluster M10 lies near Gamma Scuti.

Sagittarius contains more Messier objects than any other constellation. We have an embarrassment of riches here! Because there are so many glorious star-fields and rich areas in Sagittarius, the whole region can be rather difficult to sort out. From latitudes such as that of Britain, there is the additional complication that the altitude is always very low. From southern latitudes, where Sagittarius can pass overhead, there is plenty to see; I have always found that the best way to identify the various clusters and nebulæ is to take them 'one by one', using the nearest naked-eye stars as guides. Thus when Mu Sagittarii has been definitely located, it is not hard to move on to M25. M17, M21, M20 and M8, though the beginner will have to be careful not to confuse them.

Star clouds in Sagittarius (Anglo Australian Telescope Board © 1980).

129

SCORPIUS: *the Scorpion*

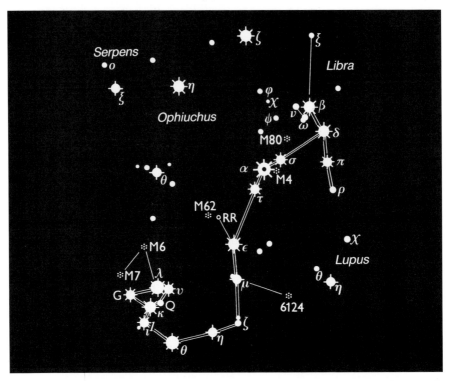

This magnificent constellation is arguably second in splendour only to Orion. Unfortunately it is too far south to be properly seen from Britain or the northern United States, but when high up it is truly impressive, and it is easy to conjure up the picture of a scorpion from the long line of bright stars, with the 'head' and the 'sting'. The leading stars are Alpha or Antares (1.0), Lambda or Shaula (1.6), Theta (1.9), Epsilon and Delta (each 2.3), Kappa (2.4), Beta (2.6), Upsilon or Lesath (2.7), Tau (2.8), Sigma and Pi (each 2.9), Iota[1] and Mu[1] (each 3.0), G (3.2) and Eta (3.3). Iota[1] may look unspectacular, but it is immensely powerful. and may be more than 150 000 times as luminous as the Sun.

Antares is a red supergiant, 7500 Sun-power and 330 light-years away. I always find it the reddest of the brilliant stars, though there is not a great deal of colour-difference between it and Betelgeux; the name 'Antares' means 'the Rival of Ares' (Mars). It is flanked to either side by Tau and Sigma, both of which rival Antares in real luminosity but are considerably further away. Antares has a greenish companion star which is a radio source, but it is not visible with binoculars.

The Scorpion's Head consists of Beta, Nu and Omega (4.0). Nu is made up of two components, easily visible with binoculars. The magnitudes are 4.3 and 6.5, and the separation is over 40 seconds of arc.

In the 'sting' (never easily seen from England and the northern United States, and not at all from Scotland) Lesath and Shaula give the impression of being a very wide pair, but they are not associated; of the two Lesath is much the more luminous and remote. These two, with Kappa, G, and Q (4.3) make up the sting. The whole of the group is in the same × 7 field, and is a lovely sight.

Zeta is a wide, easy pair; the brighter component (3.6) is orange, and the fainter (4.7) white. The separation is nearly 7 minutes of arc, but again there is no real association. The fainter member of the pair is much the more luminous, but it is over 2500 light-years away, and is actually a great deal more powerful than Antares. In the same region is another wide pair, Mu, separable with the naked eye and very easy with × 7; the magnitudes are 3.0 and 3.6. As the two components have a common motion through space they seem to be physically associated, but they are at least a light-year apart, and possibly much more.

There are plenty of variable stars in Scorpius. The brightest is RR, a Mira-type star with a period of 279 days. At maximum it may attain magnitude 5, but at minimum it falls to below 12. To find it, continue the line from Antares through Tau. RR is not far from Epsilon, which is an orange star of type K.

The Milky Way is exceptionally rich in Scorpius, and there are many clusters, both open and globular. M4, a globular, is only about 1 degree from Antares, so that it is in the same binocular field and can be found at once; it is on the fringe of naked-eye visibility, but is comparatively loose, so that a telescope is needed to resolve it. M80 is another globular, midway between Antares and Beta; I find it rather elusive with binoculars, but most observers regard it as easy. It achieved fame in 1860, when a nova appeared in it and reached the seventh magnitude, altering the whole aspect of the cluster for a few nights.

M6 and M7 are bright open clusters; both are visible with the naked-eye, not far from the Scorpion's sting, and both are easily resolved with binoculars. Another cluster within binocular range is NGC 6124 (C75), which forms a triangle with Mu and Zeta. The globular M62 is in the same field as RR Scorpii; it is about 26 000 light-years away, and is not hard to locate, though in binoculars it shows up only as a condensed blur of light.

Altogether, Scorpius has much to offer, though it is of course best seen from southern latitudes, where it can pass over the zenith.

SCULPTOR: *the Sculptor*

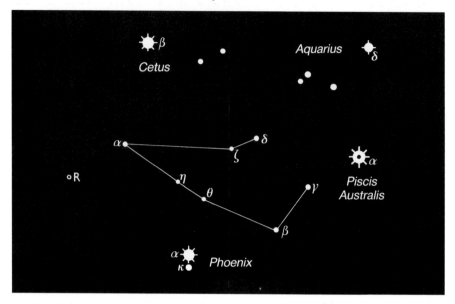

A very dim constellation; the brightest star, Alpha is only of magnitude 4.3. The best way to find it is to look some way off a line joining Beta Ceti to Alpha Phœnicis. Sculptor is always very low from British and northern United States latitudes; its only claim to fame is that it contains the south galactic pole.

SCUTUM: *the Shield*

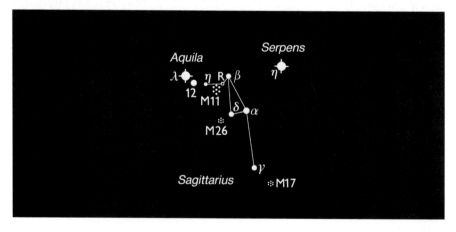

Scutum adjoins Aquila, and was formerly included in it. The only star above the fourth magnitude is Alpha (3.8), but the triangle formed by Alpha, Beta (4.2) and Delta (4.7) is not hard to locate, because Lambda and 12 Aquilæ show the way to it. Scutum is crossed by the Milky Way, and is very rich.

Delta is a short-period pulsating variable, and the prototype of its class, but as the range is only about a tenth of a magnitude it need not concern us here.

There are two objects of great interest. One is the open cluster M11, nicknamed the 'Wild Duck' because telescopically it has been likened to ducks in flight, and certainly it is fan-shaped; it contains several hundreds of stars. To find it, use Lambda and 12 Aquilæ and Eta Scuti (5.0) as guides; M11 is in the same field with a magnification of × 12 or less. The distance of the cluster is about 5500 light-years. It is visible with the naked eye, and with × 12 I have suspected individual stars; with × 20 there is undoubtedly some degree of resolution.

In the same × 7 field there is a quartet of stars, one of which is the celebrated variable R Scuti. The range is from 5.7 to 8.6 so that for almost all the time it is easy to see with binoculars. The chart given here shows suitable comparison stars.

R Scuti is one of the unusual variables classed as being of the RV Tauri type. There is a primary period of around 140 days, but neither the period nor the amplitude is constant. Generally the range is from 5 to 6, but every fourth or fifth minimum is lower, and R Scuti may become hard to see with binoculars. The star seems to be oscillating in at least two superimposed periods, and at its peak it must be at least 8000 times as luminous as the Sun.

The open cluster M26 is in the same field with Delta. I have been unable to see it with binoculars, though keener-eyed observers may be more successful.

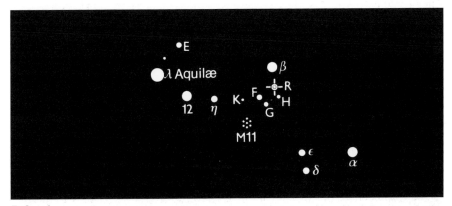

R Scuti.

The magnitudes of the comparison stars are:

α 3.9
β 4.2
η 4.8
ε 4.9
F 6.1
G 6.8
H 7.1
K 7.7

SERPENS: *the Serpent*

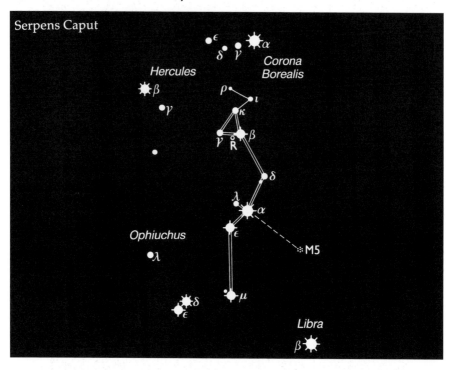

A curious constellation inasmuch as it is in two halves, separated by Ophiuchus. The Head (Caput) contains Alpha or Unukalhai (2.6), Mu (3.5) and Beta and Epsilon (each 3.7); in Cauda (the Body) are Theta (3.4) and Xi (3.5). Caput is the more obvious section, with the Serpent's Head made up of Beta, Gamma (3.8) and Kappa (4.1), arranged in a triangle and in the same × 7 field. Kappa, of type M, is decidedly red, while another orange star, Rho (4.8) is also in the field.

The Mira variable R Serpentis lies midway between Beta and Gamma. It has a period of 357 days, and at minimum is so faint that large telescopes are needed to show it, but at maximum it has been known to rise to 5.7, and is then just visible with the naked eye and very easy in binoculars.

Alpha Serpentis is a K-type orange star. Theta is a famous double; the separation is 22 seconds or arc, and the components are equal. The pair can just about be split with × 20, but certainly not with any lower magnification.

Serpens contains a prominent globular cluster, M5, which is not far below naked-eye visibility. It is just out of the × 7 field with Alpha, but Lambda (4.4) and Alpha indicate the direction in which it lies; it is close to 5 Serpentis (5.2). It is easy to see with × 7. With × 12 I find a hint of resolution in the outer parts, and this is marked with × 20. Also in the constellation is M16, the Eagle Nebula, which consists of a galactic cluster together with an emission nebula; the guide star is Gamma Scuti (4.7), and the Eagle Nebula is just over the Scutum boundary into Serpens. It is not a difficult object, though telescopes are needed to show it at all well.

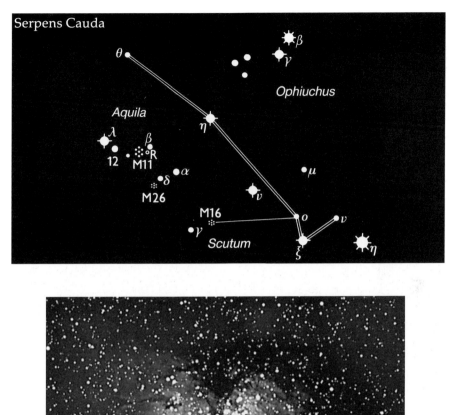

The Eagle Nebula in Serpens (Anglo Australian Telescope Board © 1980).

SEXTANS: *the Sextant*

A very barren group, containing the north galactic pole. The brightest star is Alpha (4.5), which forms a triangle with Regulus and Alpha Hydræ. Sextans contains nothing of interest to the binocular observer. It is included on the chart with Hydra.

135

TAURUS: *the Bull*

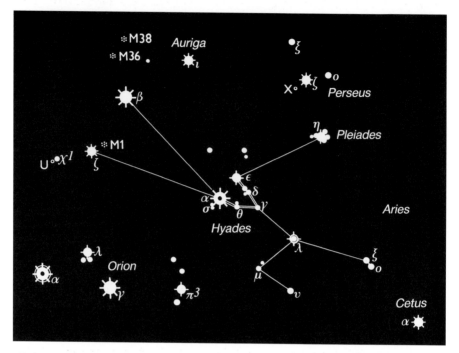

Taurus is a large and bright Zodiacal constellation, dominated by the orange-red Aldebaran (0.8) and the two open clusters of the Hyades and the Pleiades. There can be no mistaking Aldebaran, if only because it lies in line with Orion's Belt. It is 65 light-years away, and 100 times as luminous as the Sun, and with its K-type spectrum its colour is striking in binoculars. Superficially it looks very like Betelgeux, though it is not nearly so powerful.

The other leading stars of Taurus are Beta or Al Nath (1.6), Eta or Alcyone (2.9) and Zeta (3.0). Al Nath was formerly included in Auriga, as Gamma Aurigæ. The reason for its transfer is unclear, because it does seem to belong to the Auriga pattern, whereas Taurus has no obvious shape at all. Al Nath is a bluish-white star, almost 500 times as luminous as the Sun.

The Hyades cluster (C41) extends in a kind of V-formation from Aldebaran. Its distance is 130 light-years, so that Aldebaran is not a genuine member; it merely happens to lie in between the Hyades and ourselves.

The whole of the Hyades cluster is in the × 8.5 field, and almost all of it is in the × 12 field. Epsilon (3.5) and Gamma (3.6) are of type K, and off-white in binoculars. There are also several interesting pairs. Sigma consists of two dim stars, close to Aldebaran; Delta (3.8), makes up a pair with the fainter star 64 Tauri (4.8); and Theta is a naked-eye double, made up of a white star of magnitude 3.4 and a K-type orange companion of magnitude 3.8. With × 7 the colour contrast between the two Thetas is noticeable, and with × 12 or × 20 it is striking. The two are not closely associated, because the white star is the closer to us by some 15 light-years, though presumably both condensed out of the same nebula which produced all the rest of the Hyades.

Prominent though they are, the Hyades pale in comparison with the Pleiades or Seven Sisters, led by Alcyone. Here the distance is 410 light-years, and all the main stars are hot and white. The whole cluster is in the same binocular field, even with × 20, and the shape is very characteristic. The leaders, after Alcyone, are Atlas (3.6), Electra (3.7), Maia (3.9), Merope (4.2), Taygete (4.3), Pleione (5.1, but rather variable), Celæno (5.4) and Asterope (5.6); Atlas and Pleione are close together, but × 7 will separate them easily. All these stars are concentrated in a field only a little over a degree in diameter. There is nebulosity spread through the cluster, though photography is needed to show it properly.

Even more important to astronomers is the Crab Nebula, M1, the wreck of the brilliant supernova seen in 1054 (it is the only supernova remnant listed by Messier). Photographs show it as an expanding, immensely complex gas-cloud, and in its centre there is a pulsar, one of the few to have been identified optically.

Most books say that the Crab is too faint to be seen with binoculars. This is not true. With × 20 it is not hard to locate, somewhat to the north-west of Zeta and in the same field with it; I can glimpse it with × 12 and suspect it with × 8.5, though I have never been able to see it with any lower magnification. It is certainly worth finding, because it is one of the most unusual objects in the sky.

Lambda Tauri is an Algol-type eclipsing binary, with a range of from 3.3 to 4.2 and a period of 3.9 days. It may be found by using the V of the Hyades as a kind of arrow-head; with × 8.5 it is just in the field with Gamma. Useful comparison stars are Gamma (3.6), Mu (4.3), Omicron (3.6) and Xi (3.7); Omicron has a G-type spectrum, but when seen through binoculars seems to me rather more orange than might be expected. Lambda is considerably more luminous than Algol, but appears fainter because it is further away.

The Pleiades in Taurus (Royal Observatory Edinburgh © 1985).

TELESCOPIUM: *the Telescope*

A small, obscure group. Its two main stars Alpha (3.3) and Zeta (4.1) make up a small triangle with Theta Aræ, and I have included Telescopium in the map with Ara. There is nothing here of interest to the binocular observer.

TRIANGULUM: *the Triangle*

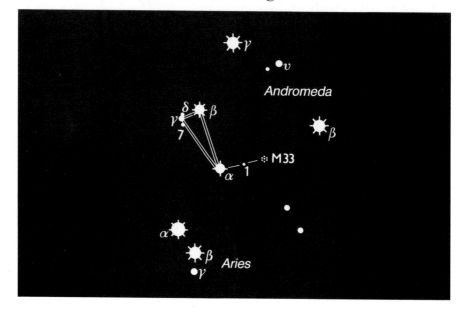

This is one of the few constellations to resemble the object after which it is named. The three main stars, Beta (3.0), Alpha (3.4) and Gamma (4.0) do indeed make up a triangle, midway between Alpha Arietis and Gamma Andromedæ. Gamma forms a pretty little group with Delta (4.6) and 7 (5.3).

Triangulum is graced by the presence of the spiral galaxy M33, a member of the Local Group at a distance of 2 300 000 light-years – only slightly further away than the Andromeda Spiral. However, M33 is a much smaller and much looser system. Some observers can see it with the naked-eye under ideal conditions, though I certainly cannot.

With × 7 binoculars it is not difficult to locate. It is in the same field with Alpha; look past the fainter star 1 Trianguli, and at about an equal distance beyond it, in the direction of Beta Andromedæ, you will see M33 as a fairly large, dim haze. I find it much easier with × 8.5 than with × 7. With × 12 it is just in the field with Alpha; with × 20 it is out of the Alpha field, but much more prominent. Oddly enough it can be a tricky object to locate in a telescope, because its surface brightness is so low, and there is no chance of seeing the spiral shape except with a very large instrument.

TRIANGULUM AUSTRALE:

the Southern triangle

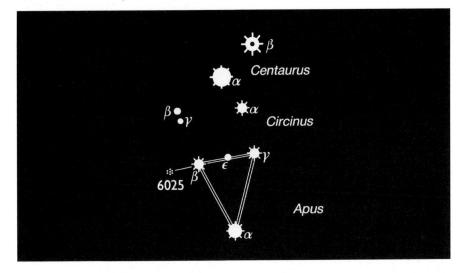

This is easy to identify. The leading stars, Alpha (1.9), Beta (2.8) and Gamma (2.9) form a triangle adjoining Alpha and Beta Centauri, the Pointers to the Southern Cross. Alpha Trianguli Australe is orange, and is bright enough to be unmistakable; its spectral type is K. Epsilon (4.1) lies midway between Gamma and Beta. Following on this line past Beta leads to the bright open cluster NGC 6025 (C95), which is easy with × 7 and is on the fringe of naked-eye visibility.

TUCANA: *the Toucan*

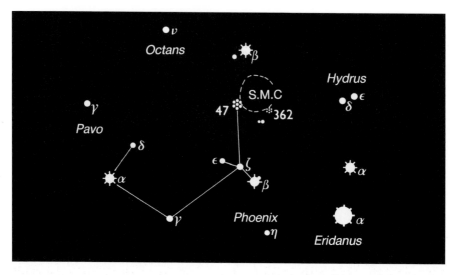

Tucana is the faintest of the Southern Birds; its leading stars are Alpha (2.9) and Beta (combined magnitude 3.7). However, Tucana contains the Small Cloud of Magellan and two superb globular clusters. Beta is a wide pair; it makes up a lovely spectacle in binoculars and is in a rich field.

The Small Cloud is very obvious with the naked eye, and binoculars show it well, though admittedly it cannot rival the splendour of the Large Cloud; it has no well-defined shape, but is easy to resolve at least in part. It is of course an external system, well over 150 000 light-years away, and it is sheer coincidence that it lies almost behind the two globulars, 47 Tucanæ and NGC 362.

47 Tucanæ (NGC 104, C106) is inferior only to Omega Centauri. It is a naked-eye object, and binoculars show it excellently; its actual surface brightness is much greater than that of the Small Cloud. Any magnification of over × 7 will start to resolve it, and with × 20 many stars in it can be seen. Its distance is about 15 000 light-years, so that it is one of the closest of the globulars. NGC 362 (C104) is hardly inferior to 47, though it is less condensed and less easy to resolve.

URSA MAJOR: *the Great Bear*

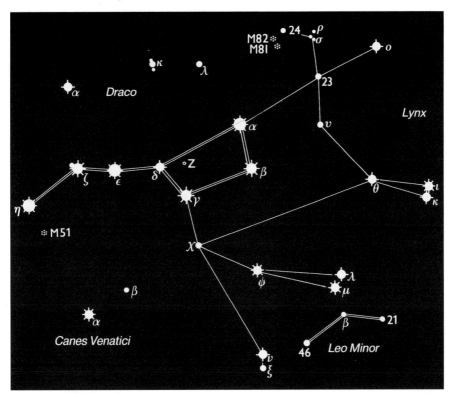

The most celebrated of all the northern constellations is distinctive even though it contains no star as bright as the first magnitude. The main pattern is often nicknamed the Plough, King Charles' Wain, or (in America) the Big Dipper, Ursa Minor being the Little Dipper. The leading stars are Epsilon or Alioth and Alpha or Dubhe (each 1.8), Eta or Alkaid (1.9), Zeta or Mizar (2.1), Beta or Merak and Gamma or Phad (each 2.4), Psi, Mu and Iota (each 3.1), Theta (3.2), Delta or Megrez (3.3), Omicron and Lambda (each 3.4), Nu (3.5) and Kappa (3.6). Mu and Lambda, in the same field even with × 20, provide an excellent colour contrast; Lambda is white, while Mu, with its M-type spectrum, is very red.

Ancient astronomers rated Megrez, in the Plough, as equal to the other stars in the pattern but it is now almost a magnitude fainter than Phad. Either there has been a real fading, or (more probably) there has been an error in recording or interpretation, though it is true that Megrez has been suspected of slight variability in modern times.

Mizar is the celebrated double star. Its companion, Alcor or 80 Ursæ Majoris, is of magnitude 4.0, and is easily visible with the naked eye; the angular distance between the two is 11.8 minutes of arc, so that in binoculars they make a noble pair. Between them – or, more accurately, forming a triangle with them – is an eighth-magnitude star, named Sidus Ludovicianum in 1723 by courtiers of the Emperor Ludwig V, who believed that it had appeared sud-

URSA MAJOR

denly. I can glimpse it with ×12 and it is easy with × 20, though I find it extremely difficult with × 7. There is a minor mystery here. Centuries ago people regarded Alcor as a naked-eye test, whereas today it is perfectly obvious, under reasonable conditions, to anyone with normal sight. Mizar itself is a splendid telescopic double, with rather unequal components, but the separation (14.5 seconds of arc) is not enough for the pair to be split with binoculars.

Dubhe and Merak are the Pointers to the Pole Star, and are in the same field with magnifications up to × 12. They are different in colour; Dubhe is an orange K-type star, whereas Merak, like all the other members of the Plough, is white. Incidentally, Dubhe and Alkaid are moving across the sky in a direction opposite to that of the remaining five stars, so that over a sufficiently long period the Plough will lose its familiar shape.

Not many variables lie in the constellation, but there is an M-type semi-regular star, Z Ursæ Majoris, in the × 12 field with Megrez. I can never see much colour in it, though it looks slightly off-white. The range is from 6.8 to 9.1, and there is a rough period of about 198 days.

There are two particularly interesting galaxies in Ursa Major: the spiral M81 and the irregular M82, which is a well-known radio source. Each is about 8 500 000 light-years away, and they are certainly associated. It has been said that with × 20 binoculars they make a striking pair.

To find M81, first identify Upsilon (3.8) and 23 (3.7). Using × 7, 23 is in the same field with the little trio made up Sigma^1, Sigma^2 and Rho, all of which are around magnitude 5. In turn the trio is in the same field as 24 Ursæ Majoris (4.6), and the two galaxies are in the field with 24. I can suspect M81 with × 7; it is detectable with × 12 and not difficult with × 20 but in spite of all my efforts, I have never been able to see M82 without a telescope. Other observers may be more fortunate. Ursa Major is rich in galaxies, but all the rest are below binocular range.

The Plough is circumpolar from Britain. It can be seen from the northern parts of Australia and South Africa, but not all of it rises from Sydney or Cape Town, and from Wellington in New Zealand it is never visible at all.

URSA MINOR: *the Little Bear*

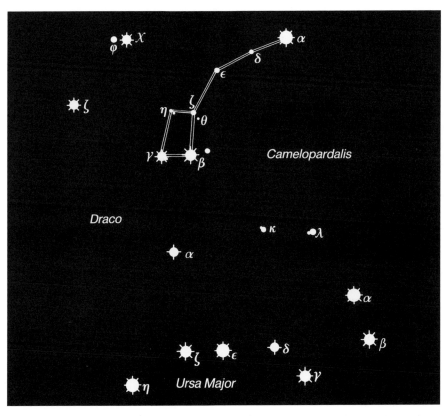

The presence of Polaris, within one degree of the north celestial pole, gives Ursa Minor its importance. The pattern of stars extending in the general direction of Mizar is unmistakable, and is slightly like a dim, distorted version of Ursa Major. Polaris, magnitude 2.0 (very slightly variable) is 680 light-years away, and 6000 times as luminous as the Sun. It has an F-type spectrum, but to me it appears pure white. The other leading stars of Ursa Minor are Beta or Kocab (2.1) and Gamma (3.0), known as the Guardians of the Pole; Kocab is strangly orange, with a K-type spectrum, and makes a good contrast with Polaris. The Guardians are in the same × 12 field.

The rest of the stars in the main pattern are Delta (4.4), Epsilon (4.2) Zeta (4.3) and Eta (4.9); moonlight will drown them. Zeta is in the same field with the reddish, K-type Theta (5.3), which is easy to see with × 7 or higher magnifications.

VELA: *the Sails*

This is part of the old Argo. It adjoins Carina, and two of its stars, Kappa and Delta, make up the False Cross with Iota and Epsilon Carinæ; for comments about the False Cross, see the section dealing with Carina. The leading stars of Vela are Gamma (1.8), Delta (2.0), Lambda (2.2), Kappa (2.5), Mu (2.7), N (3.1), Phi (3.5), and Psi and Omicron (each 3.6).

Gamma Velorum is an unusual star. It is very hot and unstable, and is of the Wolf–Rayet type (spectrum W). Stars of this kind are ejecting material, and it has even been suggested that Gamma Velorum may become a super-nova in the future, though probably not for a very long time yet. It is a wide double; the companion is of magnitude 4.8, and the separation 41 seconds of arc, so that I can see it with × 20. Also in the field with Gamma is V Velorum, a Cepheid with a range of from 7.2 to 7.9, and a period of 4.4 days. N Velorum, near the False Cross (actually in the × 7 field with Iota Carinæ) is reddish, with a K-type spectrum. It has been suspected of variability, but at present, at least, the magnitude seems to be steady at 3.1.

In the same × 12 field with Delta is the attractive little cluster IC 2391 (C85); the brightest star in it is o Velorum (3.6). It is easily visible with the naked eye, and binoculars give it a vaguely cruciform appearance. Another naked-eye cluster is NGC 2547, close to Gamma.

Lambda is a red K-type star over 5000 times as luminous as the Sun; Binoculars show the colour well. Vela is crossed by the Milky Way, though it is not so rich as Carina.

VIRGO: *the Virgin*

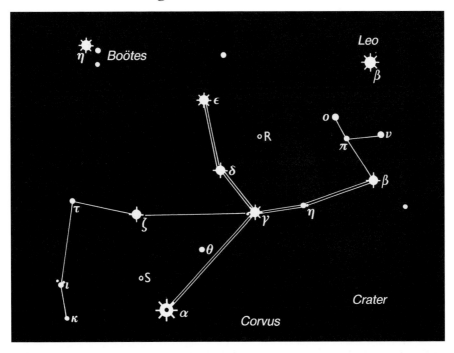

Virgo is in the Zodiac, and is one of the largest constellations in the sky. The brightest star is Spica (1.0). Next come Gamma and Epsilon (each 2.8), Zeta and Delta (each 3.4) and Beta (3.6).

Spica, over 2000 times as luminous as the Sun, is bluish-white. It is actually a binary, but the two components are so close together that they cannot be separated with ordinary telescopes. Gamma Virginis or Arich is also a binary; the identical components can be split with almost any telescope, though not with binoculars (they are much less wide apart now than they used to be 30 years ago, and the separation is decreasing steadily).

Virgo makes up a characteristic Y-shape. Delta is of spectral type M; the colour is obvious with binoculars, and with × 7 Delta and Gamma are in the same field. Eta (3.9), in the Y, was ranked as of the second magnitude in ancient times, but as it is an A-type star it is not likely that there has been any real change in it. Nu, in the field with Beta, is of magnitude 3.9 and is another M-type red star. It is only 93 light-years away, closer than most of the red giants.

The 'bowl' of the Y is crowded with faint galaxies. None of these can be identified with binoculars, but the whole area is rich enough to merit sweeping with a low power.

VOLANS: *the Flying Fish*

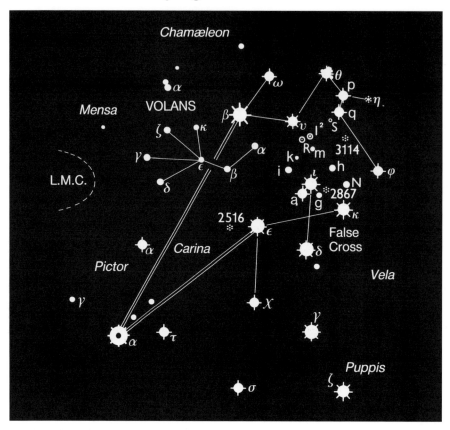

Originally Piscis Volans. It intrudes into Carina, and seems to have no justification for separate identity; it lies between Canopus and Beta Carinæ. Its brightest stars are Gamma (3.6). Beta (3.8), Zeta (3.9) and Delta (4.0). Gamma is an easy telescopic double, but not wide enough to be split with binoculars. Alpha, Beta and Epsilon (4.4) are in the × 7 field with Beta Carinæ.

VULPECULA: *the Fox*

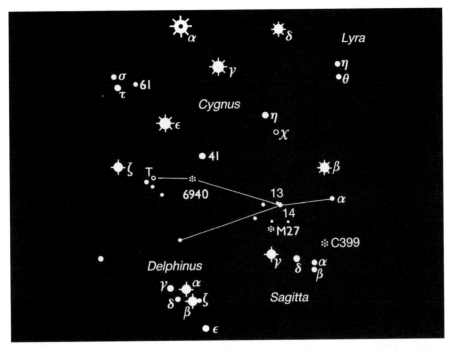

This is a very obscure group; the brightest star, Alpha, is only of magnitude 4.4. It lies between Sagitta and Cygnus.

The very loose cluster C399 can be found by using Alpha and Beta Sagittæ as pointers. The line from them passes through 9 Sagittæ and then comes to C399, which includes the faint stars 4 and 5 Vulpeculæ. The cluster is identifiable with the naked eye, and is clear with binoculars. Another open cluster is NGC 6940, in the same × 7 field with Epsilon and 41 Cygni; with × 20 it is still in the field with 41. It has very low surface brightness, and is best seen with low powers as an oval blur.

Much the most interesting object in Vulpecula is M27, the Dumbbell Nebula, which is a fine planetary. It lies 3 degrees north of Gamma Sagittæ, and this is the best way to find it; it is between Gamma Sagittæ and the dim 14 Vulpeculæ (5.7). I find it difficult with × 7, fairly easy with × 8.5 and very easy with any higher magnification. With most binoculars it is in the same field with Gamma Sagittæ, but is just out of the field with × 20. It appears as a dim but quite unmistakable blur. The distance is 975 light-years, and the present diameter is of the order of 2½ light-years. Of course, telescopes are needed to show it properly.

Finally there is the Cepheid variable T Vulpeculæ, with a range of from 5.4 to 6.1 and a period of 4.4 days. It lies near Zeta Cygni, close to a pair of stars, 31 Vulpeculæ (5.1) and 32 Vulpeculæ (5.5, of type M and obviously reddish). The variable is always visible in binoculars, though it has no features of special interest.

8

The Sun and its Eclipses

Exploring the night sky with binoculars is immensely rewarding. There is endless variety, and there is always something new to see. When we come to the daytime sky, the situation is much less attractive. The Moon can be seen often enough, but is better observed after dark, and of the stars and planets only Venus is at all easy to locate. This leaves us with the Sun, our own particular star.

As we have known for centuries, the Sun is a huge globe of incandescent gas, big enough to swallow well over a million bodies the volume of the Earth. The surface temperature is not far short of 6000 degrees Celsius, which is fairly mild for a star but very hot indeed by everyday standards. And herein lies the danger. To look straight at the Sun through any telescope or binoculars is to invite blindness. You will simply focus the intense heat on to your eyeball, with disastrous results. This is not mere alarmism; I have known it happen. If possible, binoculars are even more dangerous than telescopes, because both eyes are involved instead of only one. If anyone has doubts, simply remember the old Boy Scout method of starting a camp fire by using a magnifying glass to focus the Sun's heat on to dry material.

The real trap is that some manufacturers, who should certainly know better, provide dark 'sun-caps' which, it is claimed, can be fitted over the eye-end of a telescope so that the Sun can be observed direct. Actually, the procedure is terrifyingly dangerous. First, no dark filter can provide proper protection. Secondly, any filter is liable to shatter abruptly under the intense heat, giving the observer no time to remove his eye from the danger-zone.

I am sorry to be so emphatic about this, because I have given the warning so many times that I have been accused of being tedious, but it is really necessary. It is not even safe to stare at the Sun with the naked eye. As for using binoculars or a telescope for direct viewing, I have only one word of advice: **don't**.

There is a completely safe way to use a telescope for observing the Sun: by projection. Simply aim the instrument toward the Sun, keeping your eye well away from the eyepiece, and project the image on to a white screen or card held at a suitable distance. Any sunspots or other features on the disk will be seen with no difficulty at all. The same method can be used for binoculars, though a mounting for them is essential; trying to hand-hold is, at best, extremely difficult. Sunspots can be seen by this method, though personally I regard it as very unsatisfactory.

Sunspots are darker patches on the Sun's bright face or photosphere. They are not genuinely black; they are some 2000 degrees cooler than the surrounding regions, so that they appear dark by contrast, though if a spot could be seen shining on its own the surface brilliancy would be greater than that of an arc-lamp. They are places where the Sun's lines of magnetic force break through to the surface from below, cooling it. A large spot has a central dark umbra and a lighter surrounding penumbra, often irregular in shape; there may be many umbræ in one penumbral mass, and spots generally, though not invariably, appear in groups. The lifetime of a large group may amount to several months. Associated with them are bright streaks known as faculæ (Latin, 'torches'), which may be regarded as high-level luminous clouds. Because the Sun rotates on its axis, in a period of rather less than a month, the spots are carried slowly from one side of the disk to the other.

In some respects the Sun must be classed as a variable star. There is a definite cycle of activity, with maxima every eleven years or so; at these times spot-groups are plentiful, whereas at solar minimum the disk may be blank for many consecutive days. The maximum of the last cycle was in 1990. The peak of the current cycle is 2001, though the cycle is by no means regular. The mid-1990s were years of low activity, and during 1995, for example, my records showed that there were no spots at all for many consecutive days.

I do not propose to go into further detail here, because to study the Sun it is really rather pointless to use binoculars at all. Neither are binoculars of much help in making records of auroræ or polar lights, which are caused by electrified particles from the Sun entering the Earth's upper atmosphere and making it glow. Auroræ are best seen from high latitudes; thus from Scotland or North Norway they are common, but from South England or the New York area they are rare even when the Sun is at its most active, and they are seldom seen from South Africa or Australia, though some good displays occur over the southern part of New Zealand. The auroral patterns change quickly, and cover wide areas of the sky, so that the best views of them are obtained with the naked eye.

The Moon revolves round the Earth, while the Earth revolves round the Sun. Therefore there must be times when the three bodies line up; the Moon blocks out the bright disk of the Sun, and produces a solar eclipse. This does not happen every month, because the Moon's orbit is appreciably tilted, and on most occasions the new Moon passes unseen above or below the Sun in the sky. (It does not need much careful thinking to realize that a solar eclipse can take place only when the Moon is new!)

By sheer coincidence the Sun and the Moon appear virtually the same size in the sky; the Sun's diameter is 400 times that of the Moon, but it is also 400 times further away from us. If the lining-up is perfect, the eclipse is total, and the sight is magnificent; the Sun's atmosphere flashes into view, and for a few fleeting minutes we can see the pearly corona and the masses of red hydrogen gas which we call prominences. However, no total eclipse can last for as long as eight minutes, and most are shorter. Because the Moon's shadow is only just long enough to touch the surface of the Earth, total eclipses are rare as seen from any one location. A list of solar eclipses up to 2010 is given on page 195.

Binoculars are of limited value for observing eclipses, and it is vitally important to remember that the slightest segment of the bright disk of the Sun is dangerous. I do not recommend using binoculars at all, because totality ends with surprising suddenness. If the Sun is not completely covered, the eclipse is partial, and it is possible to use binoculars to project the image, but I am by no means enthusiastic.

All in all, the Sun is of very limited interest to the binocular-user. If you do make up your mind to try sunspot-watching by the projection method, you may well have some success, but remember always to take the greatest care. A single careless mistake can have tragic consequences.

9

Lunar Landscapes

If the Sun is a disappointment to the binocular-owner, the Moon more than compensates. There is a wealth of detail to be seen, and it is easy enough to learn one's way around.

The Moon is officially classed as the Earth's satellite, though it is probably better to regard it as a companion planet. It is not only smaller than the Earth but is also less dense with only 1/81 of the Earth's mass. This has important consequences. The Earth, with its strong pull, has been able to retain a thick atmosphere, but the Moon's gravity is much weaker, and it has been unable to hold on to any atmosphere it may once have had. To all intents and purposes, the Moon is an airless world. This means that it must also be waterless – and lifeless.

Even the naked eye will show the broad grey plains which we still mis-call seas, even though they are bone-dry. There has never been any water in them, though long ago they must have been oceans of lava. There are mountains, some of which tower to well over 6000 metres above the mean surface level; there are ridges, valleys and isolated peaks, and there are thousands upon thousands of the walled circular formations which we know as craters. To be more precise, a typical lunar crater is a depression, with ramparts which may rise high above the deepest part of the floor but only to a modest altitude above the surrounding landscape. Many of the craters have central mountains or mountain-groups, and some of the largest formations are well over 200 km in diameter, so that they dwarf any craters to be found on Earth.

Casual viewing will show one interesting fact: the lunar features seem to remain in virtually the same positions on the disk. Consider, for example, the well-marked grey plain of the Mare Crisium (Sea of Crises). It lies near the edge of the Moon's face, and can never appear anywhere else, so that to all intents and purposes it is fixed in position. This is because the Moon takes 27.3 days to spin once on its axis, and exactly the same time to complete one orbit. The result is that the same part of the Moon is turned toward us all the time, while another part is permanently hidden. Before 1959, when the Russians sent an unmanned space-craft on a 'round trip', we knew nothing positive about the far side of the Moon.

Coincidence? Certainly not. Tidal effects over the ages have led to this state of affairs, known technically as captured or synchronous rotation. In the early stages of the Solar System the Moon was closer to the Earth than it is now, and it was viscous, rather than rigid as it is today. The Earth raised powerful

151

N

S

Major features which may be seen with good binoculars.

tides in its globe, and as the Moon rotated it had to fight against a force which was tending to keep a tidal bulge turned earthward; the situation was not unlike that of a cycle-wheel rotating in spite of the slowing-down effects of two brake shoes. Eventually the Moon's rotation has been slowed down so much that relative to the Earth it had stopped altogether. Note that although the Moon keeps the same face turned to the Earth, it does not keep the same face turned toward the Sun, so that day and night conditions are the same on the far side as they are on the side which we can see – apart from the fact that on the far side the Earth will never be above the horizon. (There are various effects known as librations which complicate matters somewhat, and altogether we can examine a grand total of 59 per cent of the lunar surface from our vantage point on Earth, though never more than 50 per cent at any one moment. However, this is not important to the binocular-user.)

There are still some people who fondly believe that the Moon does not spin at all, but a little thought will show that this is wrong; a non-rotating Moon would show all its surface to us over the course of a single orbit The best analogy I can think of is to picture a man walking round a chair, turning so as to keep his face turned chairward. Anyone sitting on the chair will never see the back of the walker's neck. This is how the Moon behaves with respect to the Earth, and, incidentally, all the other large satellites of other planets have similarly captured rotations.

The Moon's distance from us is, on average, 384 400 km, roughly equivalent to a journey ten times round the Earth's equator. Cosmically it is on our doorstep, and we can see it in real detail, particularly as there are no clouds or mists to veil its surface. It sends us practically no heat, and so there is no danger in looking straight at it with optical equipment. You may dazzle yourself for a few moments, but you cannot hurt your eyes. It is rather surprising to find that the lunar surface is not as shiny as a mirror; indeed, it reflects only about seven per cent of the sunlight falling upon it, so that the Moon's rocks are dark.

Obviously, our view of the Moon is reversed as seen from the two hemispheres of the Earth. It is pointless to give two sets of charts, so I have oriented the pictures here with north at the top and west to the left. It is easy enough for southern-hemisphere observers to invert the charts and photographs, and luckily the various features are so distinctive that there can be no confusion.

First and foremost there are the seas (Latin, maria). Originally they were believed to be water-filled, and we still retain the old, romantic names. The most important of the maria are:

Mare Crisium (Sea of Crises), a relatively small, separated sea with well marked boundaries, coming into view soon after new Moon and being lost soon after full.

Mare Fœcunditatis (Sea of Fertility), *Mare Nectaris* (Sea of Nectar) and *Mare Tranquillitatis* (Sea of Tranquillity), which make up a connected system on the eastern or evening half-Moon. The *Mare Nectaris* is reasonably regular, and is joined on to the Mare Tranquillitatis by a narrow 'neck'. The Mare Tranquillitatis was the site of the first manned landing on the Moon, by Neil Armstrong and Edwin Aldrin in July 1969.

Mare Serenitatis (Sea of Serenity) leads off from the Mare Tranquillitatis. It is well-marked and regular in outline, with few major craters on its floor. Extending from it to the north are the *Lacus Somniorum* (Lake of the Dreamers) and *Mare Frigoris* (Sea of Cold); the Mare Frigoris is irregular, and stretches from the eastern side of the Moon into the western hemisphere, so that only part of it has come into view at the time of half-phase.

Mare Imbrium (Sea of Showers) is the most prominent of all the maria. It comes into view after half-phase, when the Moon is waxing, and remains visible until the Moon is becoming a waning crescent again in the morning sky. Parts of its border are formed by mountain ranges, notably the Apennines and the Alps, and there are some large craters on its floor, notably the 80-km Archimedes. The Mare Imbrium is joined on to the huge but less well-defined *Oceanus Procellarum* (Ocean of Storms), which extends almost to the western limb of the Moon.

The *Mare Nubium* (Sea of Clouds) is also joined on to the main system, and occupies a large part of the south-west quadrant of the surface. The *Mare Humorum* (Sea of Humours) is well-defined, and in form is a little like the Mare Nectaris, since it is joined on to the Mare Nubium via a narrow 'neck'; we can still see parts of the old boundary which must once have divided the two. Extending eastward from the Mare Nubium are the much smaller *Sinus Medii* (Central Bay), *Sinus Æstuum* (Bay of Heats), and finally the *Mare Vaporum* (Sea of Vapours), which is crossed by the Moon's central meridian and lies close to the lunar equator. It is notable that all these maria lie entirely upon the Earth-turned side of the Moon. There are various smaller maria near the limb, notably the *Mare Australe* (Southern Sea) in the south-east and the *Mare Humboldtianum* (Humboldt's Sea) in the north-east, but these are not well seen, because they are so foreshortened. Thanks to the space-craft and the Apollo astronauts, we now have detailed maps of the entire lunar surface, and it is true that on the far side, never visible from Earth, there are no seas at all comparable with the Mare Imbrium or Mare Serenitatis, though there is one sea, the *Mare Orientale* (Eastern Sea) which does extend on to the hidden regions. Piquantly, the Mare Orientale is to be seen on the Moon's western limb. I named it myself, because I discovered it while charting the libration zones in 1946; at that time 'east' and 'west' were used in the opposite sense to modern charts!

Generally speaking, the Moon's mountain ranges are different in nature from our own Himalayas or Andes. The main chains form the boundaries of the regular maria. The Apennines, bordering the Mare Imbrium, are much the most spectacular, and have peaks rising to over 4500 metres in places. The Lunar Alps, also bordering the Mare Imbrium, are cut through by a magnificent valley which is broad enough to be well seen with binoculars. There are many other cracklike features known as clefts or rills (German, *rilles*), though some of them turn out to be made up of chains of craters which have run together, often with the destruction of their connecting walls.

The craters dominate the whole lunar scene. Perhaps the term is misleading; 'walled plains' would be better. In size they range from vast enclosures over 200 km in diameter down to tiny pits beyond the range of any Earth-based telescope. Basically they are circular, but they break into each other and distort each other, so that in many cases they are almost obliterated. The walls may be high and continuous, often terraced; the floors are deeply sunken (over 3000 metres below the rampart crests in some cases), and central mountain structures are common, though there are many craters which lack them. In profile they are more like shallow saucers than steep-sided mine-shafts, and the slopes of the walls are comparatively gentle. For example, Ptolemæus, near the apparent centre of the Moon's face, has a diameter of nearly 150 km, but the walls nowhere rise to more than 1250 metres, and in places they are discontinuous. The craters have been named after famous personalities, generally astronomers. Tycho, Copernicus, Kepler, Newton and others are all commemorated – though some curious characters are to be found here and there; for example Julius Cæsar has his own crater, because of his connection with calendar reform, and there are even a couple of Olympians, Atlas and Hercules. One crater has the surprising name of Hell, honouring Maximilian Hell, a Hungarian astronomer of two centuries ago.

Craters near the Moon's limb are badly foreshortened, and even if genuinely circular, as most of them are, seem to be drawn out into ellipses. A good example of this effect is provided by the Mare Crisium. It seems to be elongated in a north–south direction, but in fact it is really longer in an east–west direction. Very close to the limb, it is often extremely difficult to distinguish between a crater and a ridge.

We know far more about the Moon now than we did a few decades ago, before the space-probes and the Apollo landings. The outer surface or 'regolith' is crumbly, though there are no deep, dangerous dust-drifts as a few astronomers used to believe; there is virtually no local colour, and there is complete calm, with an absence of any kind of conventional weather. There is no air to screen the surface, so that on the lunar equator the days are very hot (over 90 degrees Celsius), though the nights are bitterly cold. Since the Moon spins so slowly, daytime there lasts for almost two Earth weeks, followed by an equally long night. From the surface the sky is black, because there is no atmosphere to spread the sunlight around and make the sky blue. The Earth is magnificent in the lunar sky; I well remember the comment by Commander Eugene Cernan, the last man on the Moon (so far), who told me that during his excursion to the lunar surface, in 1972, the most impressive sight of all was that of his own world, so far away.

The Moon has no detectable overall magnetic field. The core is presumably hot, though not at so high a temperature as the centre of the Earth. No life can ever have appeared there, and analyses of the rocks brought back by the Apollo astronauts and the Russian sample-and-return probes have shown no trace of hydrated materials – that is to say materials containing H_2O in any form. Reports of ice deposits in some deep polar craters are certainly wrong. But how has the Moon's surface evolved? In particular, how were the craters formed?

Astronomers now believe that the craters were formed by meteoritic bombardment, followed by tremendous outpourings of lava from below the shattered crust. The alternative theory – that the craters themselves are of internal origin – has fallen into disfavour. One thing is certain: major changes on the Moon do not occur now, and the lunar surface has been virtually static for at least a thousand million years, probably longer. Even the so-called young craters date back to the period of primitive life on Earth, when even the ferocious dinosaurs lay in the far future.

The observer of the Moon will come up against one immediate major problem. A crater is at its most prominent when near the terminator, or boundary between the daylight and night hemispheres, when the Sun is either rising or setting over it; the walls cast long shadows across the sunken floor, and in many cases the top of the central peak is illuminated well before the lower parts of the interior are free from shadow. Under higher illumination, a crater which has been strikingly conspicuous when at the terminator may become difficult to identify at all, unless it has a floor which is either exceptionally bright or exceptionally dark. I well remember that when I first decided to look at the Moon with a small telescope, when I was aged seven, I consulted a lunar map and started to look for the majestic 148-km crater Ptolemæus. I failed to find it – because on that night the Moon was nearly full, and there were no shadows inside Ptolemæus, so that it was quite unrecognizable. Actually, full Moon is the very worst time to start finding one's way around, except insofar as the maria are concerned.

Now let us work our way through a complete 'lunation', beginning with new Moon. 'Day 1' refers to one day after new, and so on.

Days 1 and 2. The Moon is a slender crescent in the evening sky. Binoculars show little surface detail, but it often happens that the unlit side of the Moon can be seen shining quite brightly, an appearance nicknamed 'the Old Moon in the New Moon's arms'. There is no mystery about it; it is due to light reflected on to the Moon from the Earth, so that we call it the Earthshine. I have often been able to trace not only the maria, but also some of the craters, notably the very brilliant ones such as Aristarchus on the Oceanus Procellarum. Generally speaking the Earthshine is not evident after about Day 4.

Day 3. The terminator is now near the Mare Crisium, which is well seen with even × 7. The north–south diameter of the Mare is 450 km and the east–west diameter 560 km, so that the total area is greater than that of England. South of the Mare Crisium can be seen part of the Mare Fœcunditatis. With × 12 or over, some major craters can be made out, such as the 125-km Cleomedes, immediately north of the Mare Crisium, and the even larger Langrenus, Vendelinus and Petavius, bordering the Mare Fœcunditatis. Of these three, Vendelinus is much the least regular, with lower walls so that it is presumably older. Petavius is noted for the great cleft which runs from its central mountain group right up to the wall. It is not an easy binocular object, but it can be seen with × 20 provided that the binoculars are firmly mounted.

Day 4. The crescent has broadened, and the whole of the Mare Crisium is well in view. Closely west of it is the very brilliant crater Proclus, 29 km in diameter and 2400 metres deep – one of the brightest points on the whole of the Moon. Almost all the Mare Fœcunditatis can now be seen, and to the north-east there is the limb-sea Mare Humboldtianum, which is badly fore-shortened and appears as little more than a dark streak. Look also for the dark-floored crater Endymion.

Day 5. We now see the Mare Tranquillitatis, Mare Nectaris and part of the Mare Serenitatis. Adjoining the Mare Nectaris, to the west, are three great craters: Theophilus, Cyrillus and Catharina, which form a vast chain. Theophilus, the best-formed of them, is 103 km in diameter, with high, terraced walls and a prominent central mountain group; it intrudes into its neighbour, the lower-walled Cyrillus, and beyond Cyrillus lies Catharina, with much less regular walls and a rough floor with no central peak. South of the Mare Nectaris is another prominent crater, Piccolomini, which lies at the end of the Altai Scarp. It is evident that the south-eastern quadrant of the Moon is extremely rough, with vast numbers of craters of all kinds.

Day 6. The whole of the Mare Serenitatis is on view, bordered to the south by the Hæmus Mountains. There are no large craters on it, and the most conspicuous of them, Bessel, is only 19 km in diameter. On the north border, where the Mare adjoins the Lacus Somniorum, may be seen the 100-km Posidonius, with rather low walls and a floor which is crowded with fine detail, while in the higher-altitude region between the Mare Serenitatis and the eastern end of the patchier, irregular Mare Frigoris there is a noble crater-pair, Aristoteles (96 km in diameter) and Eudoxus (64 km). At the southern border of the Mare Serenitatis look for the brilliant crater Menelaus, in the Hæmus Mountains, and Plinius, which 'stands sentinel' on the strait separating the Mare Serenitatis from the Mare Tranquillitatis. Close to the crater-pair of Godin and Agrippa, × 12 binoculars should show the long, cracklike feature known as the Ariadæus Rill. Look also for Maurolycus, which is the largest member of a complex group of craters. Note that although the Theophilus chain is still fairly prominent, the craters further east, even Langrenus, have become rather hard to identify now that the Sun has risen over them and they are losing their interior shadows.

Day 7. The Moon is now at half-phase, with the whole of the eastern hemisphere visible; this is the phase known technically as First Quarter (because one-quarter of the orbit has been completed since new moon). The Mare Serenitatis is very prominent, with the brilliant Menelaus in the Hæmus range. Extending westward, the Apennines have begun to come into view. This is not the highest range on the Moon, but it is certainly the most impressive, though it will be better placed after Day 8. To the south, beyond the Sinus Medii, we come to the magnificent trio made up of Ptolemæus, Alphonsus and Arzachel. Ptolemæus is the largest of them, with a diameter of 148 km; it has low walls and a comparatively level floor. Alphonsus, its southern neighbour, is smaller but rather deeper, with a modest central mountain group; it is of special interest because it was here, in 1958, that the Russian astronomer Nikolai Kozyrev detected signs of activity in the form of a reddish patch indicating gaseous emission from beneath the surface. Events of this kind, known as T.L.P. or Transient Lunar Phenomena (a term for which I believe I was responsible) are of great importance to students of the Moon, but since they are well beyond binocular range they do not concern us here. South of Alphonsus is Arzachel, with higher, better-defined walls and a prominent central peak. South again there is another line of three

large formations: Walter, Regiomontanus and Purbach. Walter, the largest member, is 145 km across, almost the equal of Ptolemæus.

Note the tendency for lunar craters to appear in chains or groups, and also that when one formation breaks into another it is almost always the smaller crater which distorts the larger. The distribution is non-random.

Day 8. The Moon is now gibbous, that is to say over half-phase. The Apennines are coming into a better position for observation, and the eastern edge of the great Mare Imbrium has started to appear. The Ptolemæus and Walter chains are conspicuous, but craters so prominent earlier in the lunation, such as Piccolomini, are becoming hard to trace with binoculars – though Langrenus, near the eastern limb, shows up as a bright mass, and Proclus remains identifiable because of its great brilliancy. The southern highlands are on view, with large craters such as Maurolycus and Stöfler. In the north, on the border of the Mare Imbrium, the Alps are partially illuminated; look for the Alpine Valley slicing through the mountains.

In the south, some bright streaks may be detected. These come from the ray-crater Tycho, which is not yet on the sunlit part of the Moon.

Day 9. The terminator has now reached the middle of the Apennine range, and the high peaks cast long shadows on to the Mare surface below. On the Mare Imbrium itself, binoculars show three major craters making up a triangle: Archimedes, Aristillus and Autolycus. Archimedes, low-walled and with a relatively level, somewhat darkish floor, is 80 km in diameter; Aristillus and Autolycus are smaller but deeper, and Aristillus has a prominent central mountain.

On the Mare Nubium, west of Ptolemæus, look for the Straight Wall – which is badly named, because it is not straight and is not a wall; it is a 250-metre fault in the surface, and casts a pronounced shadow. It is 128 km long, and the angle of slope is probably about 40 degrees, so that it is by no means precipitous (in the future, when lunar travel is common, it will no doubt become a tourist attraction). But the main addition to the lunar scene is Tycho, only 87 km in diameter, but the major ray-centre of the Moon. The rays extend in all directions, and cover all the formations over which they pass, so that they are surface deposits; they are best seen under high illumination. South of Tycho there is the giant formation Clavius, 232 km in diameter, with a chain of smaller craters crossing its floor. Clavius is one of the largest of all the craters, and even × 7 will show it well. By now the main formations of earlier in the lunation, even Theophilus, are becoming hard to identify.

164

Day 10. The terminator has reached the western end of the Mare Imbrium. With luck, you should see the Sun catching the peaks which border the Sinus Iridum or Bay of Rainbows, which leads off from the Mare Imbrium; if conditions are suitable, the 'wall' of the Sinus Iridum is seen apparently projecting into the blackness beyond the terminator, an appearance which is often nicknamed the Jewelled Handle. It lasts for only a few hours, but it is worth looking for. Copernicus, another ray-centre, is now almost at the terminator; it is slightly larger than Tycho, with magnificently terraced walls and a central mountain group. It has been referred to as 'the Monarch of the Moon'; it lies just south of the Apennines, on the Mare Nubium. North-east of it is the 61-km crater Eratosthenes, at the extremity of the Apennines.

Immediately north of the Alps, between the Mare Imbrium and the irregular Mare Frigoris, lies the 96-km crater Plato. This is one of the most famous of all the lunar features. It appears oval (though in reality it is almost perfectly circular), and it stands out because of its iron-grey floor, which is one of the darkest areas on the whole of the Moon. Whenever Plato is on the sunlit hemisphere it is identifiable at a glance. There are other regular, dark-floored craters, but Plato is much the most conspicuous of them.

Day 11. The Jewelled Handle appearance of the Sinus Iridum is less evident, but the Bay itself is easy to find, and by now all the Mare Imbrium is on view. Tycho and Copernicus are striking, with their ray-systems becoming more and more prominent. The most notable addition to the scene is Aristarchus, on the Oceanus Procellarum. It is only 37 km in diameter and 1500 metres deep, but it is much the brightest object on the entire Moon, and on many occasions it has been mistaken for an erupting volcano. It has continuous walls, and a brilliant central peak.

In the south of the Moon, the Mare Humorum is prominent, leading off the Mare Nubium; on the western side of the 'neck' joining the two is the majestic 120-km crater Gassendi. The shadow cast by the Straight Wall is shortening, but is still easy to see.

Day 12. The Moon is nearing full phase, and over most of the disk the shadows are slight. Formations such as Maurolycus and Stöfler are lost, and the Tycho and Copernicus rays are dominant, while another ray-centre, Kepler, has come into view west of Copernicus. Kepler is much smaller than the two main ray-craters, and its system of bright streaks is not so conspicuous, but binoculars give a good view.

Day 13. Almost full, and by now the rays, particularly those from Tycho, are so prominent that they tend to obscure much of the detail formerly visible. Aristarchus stands out because of its brilliance, Plato because of its darkness. It is interesting to compare Aristarchus with Proclus on the edge of the Mare Crisium; Aristarchus is considerably the brighter, but Proclus continues to be easily seen.

Day 14. The Moon's synodic period (that is to say, the interval between one new Moon and the next) is 29 days 12½ hours; therefore, full Moon occurs between 14 and 15 days after new. For a few hours the edge of the disk appears circular, and there are almost no shadows anywhere, so that all that can be seen is a medley of bright and dark markings, with the great seas and ray-systems dominant. Note Grimaldi, 193 km in diameter, close to the western limb; its floor is the darkest on the Moon, so that it cannot be mistaken. Closely north of it is Riccioli, rather smaller, and with a patchier floor which also contains one very dark area. The Straight Wall has been lost; after full moon the sunlight strikes its face, so that it appears as a bright line instead of a black one.

Within a few hours after full, the terminator becomes evident once more, this time on the eastern side; but it takes time for the ray-systems to fade appreciably and the shadows to return.

Days 15 to 20. The Moon is now waning, and features such as the Mare Crisium, Mare Fœcunditatis, Mare Tranquillitatis and Mare Serenitatis are lost in turn; as the terminator reaches the Theophilus chain, the great walled formations become prominent for a while before being engulfed in the lunar night. Everything is happening in the reverse order, though on the western hemisphere Aristarchus, Plato and Grimaldi continue to stand out.

Day 21. We are back to half-moon, but this time it is the western hemisphere which is on view – and since this contains much more Mare surface than the eastern, the morning half-moon is considerably the less brilliant. Ptolemæus is back on the terminator; the Straight Wall is an easily-identified bright line; Tycho is still brilliant even though its rays are departing from the scene. Clavius is prominent in the far south, and Plato has yet to make its exit.

Days 22 and 23. The crescent narrows; the Apennines are lost, as is the Mare Imbrium. By Day 23 the Sinus Iridum is back at the terminator, though since it is the eastern part which vanishes first there is no Jewelled Handle. Copernicus is entering its period of night; Clavius has gone, and so has Plato, but Aristarchus remains as a glittering point against the greyness of the Oceanus Procellarum, and Kepler can still be made out.

Days 24 to 28. The details are disappearing as the Moon nears the Sun, and is above the horizon only briefly against a brightening sky. Aristarchus vanishes; probably the last feature to be identifiable is Grimaldi, which is lost only shortly before new Moon, and shows up defiantly as a dark patch close to the limb. On **Day 29** the Moon is new once more, to reappear soon afterwards at the start of a fresh lunation.

Eclipses of the Moon are quite different from those of the Sun. If the Earth, Sun and Moon line up, with the Earth in the mid-position, the Moon passes into the Earth's shadow, and its supply of direct sunlight is cut off. The Moon does not (usually) vanish altogether, because some of the Sun's rays are refracted on to its surface by way of the layers of atmosphere surrounding the Earth, but the Moon turns a dim, often coppery colour until it emerges from the shadow. Lunar eclipses may be either total or partial.

The main cone of the Earth's shadow is known as the umbra. On average it is about 1 340 000 km long, so that at the mean distance of the Moon it has a diameter of around 9000 km. Unlike a solar eclipse, a lunar eclipse is a gradual, protracted affair; totality may last for anything up to an hour and three-quarters. Obviously, the Moon can be eclipsed only when it is full.

Because the Sun is a disk, and not a point of light, the umbral cone is bordered by an area of partial shadow, called the penumbra. This also causes a dimming of the Moon, though not to so great an extent. Since the Moon has to pass through the penumbra before entering the umbra, the experienced observer will be able to detect the dimming before the main eclipse begins. Sometimes the Moon misses the umbra altogether, and passes only through the penumbra. It is often said that a penumbral eclipse is difficult to detect. I do not agree; with any binoculars I have always found it easy, and even with the naked eye I can notice it.

Binoculars give splendid views of lunar eclipses, and the colours are often glorious, with reds, greens, yellows and blues. Of course everything depends upon the state of the Earth's air, because all the light reaching the eclipsed Moon has to pass through our atmosphere. On occasion, the Moon darkens so much that it almost disappears, and it has been known to vanish completely for some time around mid-totality; in general this happens after a violent volcanic eruption or huge forest fires on Earth, so that the atmosphere is laden with dust or ash. When the air is clear, the Moon remains prominent all through an eclipse.

A total solar eclipse is visible from only a narrow track on the Earth's surface, but a lunar eclipse can be seen from any point from which the Moon is above the horizon, so that from any particular site it is lunar eclipses which are the more common. It cannot be said that they are important, but they can be beautiful, and they are well worth watching. I always feel that the best views are obtained with powerful binoculars, provided that they are well mounted; my 20×70 pair is ideal.

Now and then the Moon passes in front of a star, and hides or occults it. Because there is no air around the Moon's limb, the disappearance is abrupt, and the star snaps out like a candle-flame in the wind. Binoculars can show occultations of bright stars (or planets), but these events are surprisingly rare.

The Moon is no longer isolated. It has been reached, and no doubt men will return there within the next ten years or so. Before the middle of the 21st century there may be a flourishing lunar colony, and the relics of the first days of pioneering, notably the remains of space-craft, will have been brought into museums. There are also the 'Moon cars' taken there by Apollos 15, 16 and 17, which will remain to be collected; there is nothing to make them deteriorate, and the next explorers will have to do no more than fit them with new electric batteries and drive them off.

It has been claimed that the lunar journeys have destroyed the magic and romance of the Moon. I am sure that this is not so. Airless and waterless though it may be, it is our faithful companion in space, and its fascination never palls.

10

The Planets

Look at Venus or Jupiter, shining down far more brilliantly than any star, and it is easy to imagine that the planets are important. In fact they are not; even Jupiter, much the largest of them, has only a tiny fraction of the Sun's mass. The core of a planet is not nearly hot enough for nuclear reactions to be triggered off, and many of the smaller members of the Solar System contain a great deal of ice mixed in with rock.

Venus, Mars, Jupiter and Saturn are brilliant, while Mercury is easy to see when well placed. Uranus is on the fringe of naked-eye visibility; Neptune is quite prominent with binoculars, and only Pluto is out of range. Of the minor bodies, one asteroid – Vesta – can be detected with the naked eye at times, and there are several others which can exceed the ninth magnitude. Among the satellites, it is said that all four major attendants of Jupiter (Io, Europa, Ganymede and Callisto) can be seen with binoculars, though I always find them difficult. All other planetary satellites are too faint so far as I am concerned, though I am told that really keen-eyed observers can glimpse Titan, the senior satellite of Saturn.

I must be honest, and say that in general binoculars are of very limited use for observing planets Even × 20 is too low a magnification to show much. I once used a pair of very powerful binoculars (40 × 150) to look at Jupiter and Saturn, but the view was still very mediocre, so that planetary work is one branch of astronomy in which a telescope is more or less essential. Of course, the changing positions of the planets can be checked from night to night, but there is nothing very thrilling in this apart from the satisfaction of seeing that the planets are definitely mobile!

Occasionally a planet will be occulted by the Moon, and in such a case the immersion and emersion are not virtually instantaneous, as they are with a star. However, these events are rare, and are not easy to follow with binoculars unless Venus is involved. Even more exceptional are occultations of stars by planets. I have seen one or two, but for a good opportunity we must wait till 1 October 2044, when Venus will occult Regulus. Finally, there are occasions when one planet occults another. Since this will not happen again until 22 November 2065, when Venus will pass in front of Jupiter, I do not think that these mutual occultations need concern us here.

Planetary conjunctions are rather more frequent, but the only case in the recent past which was sufficiently far from the Sun to be well seen was that of 5 August 1989 (Mercury/Mars: 47 seconds). The planetary massing of May 2000 was drowned by the presence of the Sun.

Mercury is never striking, and there must be many people, even amateur astronomers, who have never seen it. Actually it can be bright; its maximum magnitude is −1.9, so that it surpasses even Sirius, but it is never to be seen against a dark background (except during a total solar eclipse, when the attention of observers will certainly be elsewhere). The greatest elongation from the Sun is never as much as 30 degrees, so that with the naked eye Mercury can be seen only low in the west after sunset or low in the east before sunrise. Binoculars help in locating it, but once more I must come back to the point I have stressed over and over again: never sweep around when even a segment of the Sun is above the horizon.

Frankly, the only point in looking at Mercury at all with binoculars is the sheer satisfaction of seeing it. It shows phases from new to full, and with × 20 I have suspected the phase when Mercury is in the crescent stage, but not at all clearly, and if I had not known about it I am quite sure that I would not have noticed it. Even powerful telescopes will show little on its disk and virtually all our knowledge about its surface features comes from the space-craft Mariner 10, which made three active passes of the planet in 1974–5 before its power failed. Like the Moon, Mercury is mountainous, cratered and virtually without atmosphere.

Occasionally Mercury passes in transit across the face of the Sun, and though it is then invisible with the naked eye I imagine that projection with binoculars would show it. Try by all means, taking the usual precautions, but do not be disappointed if you fail, and I am unenthusiastic. The next transits will occur on 7 May 2003, 8 November 2006 and 9 May 2016.

Venus is a very different proposition. In size and mass it is almost the Earth's equal; its mean distance from the Sun is 108 000 000 km, so that it can be seen against a dark background. At its best it can cast a shadow, and not even the most myopic observer can overlook it.

Venus is so bright partly because it is the closest of all the planets, with a minimum distance of about 100 times that of the Moon, and partly because it is so reflective. It is permanently covered with a dense, cloudy atmosphere, and its albedo or reflecting power is about 75 per cent, as against well below 10 per cent for Mercury and the Moon. It has long been known that the atmosphere consists mainly of carbon dioxide, and that the clouds contain large quantities of sulphuric acid, so that Venus is hardly an inviting world. Our information has come mainly from the various Russian and American space-craft which have been sent there. Radar mapping of the surface from probes orbiting Venus has enabled us to compile a reliable chart; there are craters, highlands, valleys, and huge volcanoes which are almost certainly active. But from Earth we can see little, and the only markings visible with telescopes are vague, cloudy patches.

Obviously, then, binoculars are of little use, but at least they will show the changing phases very clearly. When at greatest elongation, Venus appears as a half, waning to a crescent during eastern (evening) elongations and waxing when Venus is visible in the morning sky. When full, it is on the far side of the Sun and is to all intents and purposes out of view, while when at its closest to us it is new, and its dark side faces us. Transits are rare. The last occurred in 1882, and the next will not be until 2004 (it is interesting to reflect that there can surely be nobody now alive who can remember seeing a transit of Venus). According to all reports, the phenomenon is easily visible with the naked eye, so that projection with binoculars will be perfectly adequate. I can only depend upon second hand accounts; if I am still around in 2004 I will certainly watch with considerable interest.

With Venus, as with Mercury, binoculars can be used to sweep for the planet so long as the Sun is out of view, but there is a major difference: Venus is so brilliant that when at its best it can be seen with the naked eye in broad daylight provided that the observer has keen sight. There is a trap here. Find Venus, and it is tempting to turn binoculars toward it. This is safe only when the greatest care is taken, and under no circumstances sweep toward the Sun; always sweep away from it. I do not recommend the procedure, because even when Venus is located there will be little to be seen. It is true that the best telescopic observations are made in daylight, but this involves using a telescope equipped with accurate setting circles. Venus is at its brightest during the crescent stage, because the apparent diameter is then greater than at half or gibbous phase. Binoculars show the crescent easily.

Mars is even less rewarding, and binoculars will show nothing apart from a tiny, reddish disk. Though Mars can present a phase, and appear the shape of the Moon a few days from full, binoculars will not show it. There are also times when Mars looks exactly like a star. At its best it may surpass any of the other planets apart from Venus, but when faintest it becomes only slightly brighter than Polaris.

The mean distance from the Sun is 228 000 000 km, and the revolution period is 687 days. Therefore Mars comes to opposition only in alternate years; the mean synodic period (that is to say, the interval between successive oppositions) is 780 days.

When Mars is in the region of Scorpius, it is worth comparing it with its 'rival', Antares. The two look so alike that it is hard to realize that Antares is a vast supergiant star while Mars is a puny planet with a diameter only a little over half that of the Earth.

Beyond Mars come the *asteroids* – or most of them; there are some which swing away from the main swarm and may approach the Earth. Of these, Eros is the best known. When it comes closest, at only 24 000 000 km or so, it can rise to the eighth magnitude, but this happens very rarely (the last occasion was in 1975). So far as the 'regular' asteroids are concerned, there is only one, Vesta, which can ever become visible without optical aid, and even then it is a borderline case. Of the rest, there are only eight which can reach magnitude 9 or brighter: Ceres, Pallas, Iris, Juno, Hebe, Eunomia, Flora and Melpomene, whose diameters range between 1000 km for Ceres to only 109 km for Melpomene. Vesta, incidentally, is smaller than Ceres, which is much the largest of all the asteroids.

An asteroid looks exactly like a star, and the only way to identify it is to check the star-field from one night to another and pick out a starlike point which moves. Yearly almanacs give their positions, and it is quite intriguing to take a pair of binoculars and go asteroid-hunting. On the whole I find that around × 40 is the best magnification to use.

Jupiter moves round the Sun at a mean distance of 778 000 000 km in a period of almost 12 years. This means that it crosses approximately one Zodiacal constellation per year, though of course its apparent motion is not regular, and there are times when it backtracks or moves retrograde for a while before resuming its eastward march.

Jupiter cannot be mistaken because it is always very brilliant, and never drops below magnitude −2. Binoculars show it as a definite, yellowish disk, with a maximum diameter of over 50 seconds of arc. No binoculars will show any disk detail; to see the belts, the bright zones, and the Great Red Spot you need a telescope. On the other hand, it is certainly possible to make out the principal satellites.

Jupiter has an extensive family. Sixteen satellites are now known, but most are very small indeed, and we need concern ourselves here only with the four 'Galileans' – so called because they were observed by Galileo with his primitive telescope, as long ago as 1610. They have been named Io, Europa, Ganymede and Callisto.

The Galileans are remarkable bodies. The Voyager and Galileo probes have shown that they are very different; Ganymede and Callisto have icy, cratered surfaces, while Europa is icy and smooth, and Io has a red sulphur-coated surface with violently active volcanoes. However, details such as this cannot be seen from Earth, and but for the space-craft we would know nothing about them – even though the satellites are big; Ganymede and Callisto are about the size of Mercury, while Io is slightly larger than our Moon and Europa only slightly smaller.

There is conclusive evidence that people with exceptional eyesight can see the satellites, or some of them, with the naked eye. (One of them, presumably Ganymede, seems to have been noted by a Chinese named Gan De as long ago as the year BC 364, though needless to say he had no idea that it was anything but a star.) Their magnitudes range from above 5 for Ganymede down to about 6 for Callisto. This being so, they would be easy enough to see without optical aid were they not so overpowered by the brilliant light of Jupiter. They are close-in, and even Callisto can never be more than about 10 minutes of arc away from Jupiter. but it would seem that they ought to be well inside binocular range.

For this it seems reasonable to use the maximum power available, and × 20 should be adequate. I have to confess that I always find even Ganymede extremely elusive, but other observers do not agree. I suggest that you try the experiment with as many different pairs of binoculars as possible, and see which give the best results; I know several people who prefer × 12 to × 20. In any case, there is no chance of glimpsing the satellites unless the binoculars are firmly mounted. Hand-holding will not do, at least in my opinion. I may well be wrong, but I can speak only from my own experience.

Saturn, outermost of the planets known in ancient times, is smaller than Jupiter, and is also much more remote, moving round the Sun at a mean distance of 1 427 000 000 km. Its magnitude ranges from −0.3, slightly brighter than Arcturus or Alpha Centauri, down to 0.8, roughly equal to Aldebaran. The orbital period is 29.5 years, so that Saturn is a comparatively slow mover.

The magnitude depends largely upon the tilt of the rings. Telescopically, Saturn is unrivalled; the superb ring-system is in a class of its own. But though the rings are very wide, they are also wafer-thin, and recent estimates indicate that their thickness may be no more than a kilometre. When the rings are edgewise-on to the Earth, as in 1980 and 1995, they almost vanish even in large telescopes, and since the rings are more reflective than the globe these are also the times when Saturn is faintest. The Voyager probes have told us a great deal about the ring-system. The structure is much more complicated than used to be thought; instead of a few definite rings, separated by gaps, there are thousands of ringlets and narrow divisions. They are not solid, but are made up of icy particles moving round Saturn in the manner of dwarf moons.

Low-power binoculars will do no more than show that Saturn is a disk rather than a point of light. With × 12 or higher magnification, it is possible to see that there is something unusual about the shape – that is to say, when the rings are fully open, as they are around 2000–2003; but there is absolutely no chance of seeing that rings are responsible, so that to the binocular-owner I fear that Saturn is of limited interest. Titan, much the largest of the twenty satellites, is comparable in size with Mercury, and is the only planet satellite known to have a dense atmosphere, so that it is of exceptional interest. The magnitude is just below 8, but the angular distance from Saturn can never be as much as 4 minutes of arc. I have made many attempts to glimpse Titan with binoculars, but I have never succeeded; others can do better, and it is worth a try.

Uranus, the next giant, is 51 800 km in diameter, but it is a long way away – on average, 2 870 000 000 km from the Sun – and since it has an orbital period of 84 years it seems to move very slowly indeed. The magnitude can rise to 5.6, and Uranus is then distinctly visible with the naked-eye, but it looks exactly like a star, and I doubt whether you will find it without knowing its position in advance. With binoculars of × 12 or lower magnification, it appears stellar. Using × 20, I can distinguish it from a star, but not easily, and I have never been able to detect the green colour which is so obvious in a telescope.

Interesting comparisons can be made by checking the magnitude of Uranus against nearby stars, using the same methods as those of the variable star observer. There are suggestions that the brightness of Uranus may be perceptibly affected by changes in the Sun. If the Sun brightens slightly, Uranus will brighten too, and the effect will be more easily detected than with the nearer planets, which show obvious disks in binoculars. All the satellites of Uranus are smaller than our Moon, and are quite out of binocular range.

Neptune, with a magnitude of 7.7, takes 164.8 years to complete one journey round the Sun. It is slightly smaller than Uranus, but rather more massive; the colour is bluish rather than green, but the blueness cannot be detected with binoculars, and all that can be done is to identify Neptune by its slow shift in position against the starry background. Even with × 20 I cannot see any appreciable disk. Magnitude estimates may be attempted, but I doubt if they can be very accurate with binoculars, because Neptune is too faint – and as its mean distance from the Sun is nearly 5 000 000 000 km, this is hardly surprising.

I need say little about *Pluto*, discovered by Clyde Tombaugh in 1930. Between 1919 and 1999 it was closer-in than Neptune, because it has a comparatively eccentric orbit, though it has now again become 'the outermost planet'; the revolution period is 248 years. The magnitude is only 14, so that Pluto is beyond the range not only of binoculars, but also of small telescopes.

All in all, the planets are not rewarding objects for the binocular-owner, and it would be wrong to pretend otherwise. Yet there are observations to be made; it is always worth seeking out elusive Mercury, watching the changing phase of Venus, looking for the Galilean satellites of Jupiter, checking on the unusual shape of Saturn, and locating the remote, chilly Uranus and Neptune. The Sun has a very varied family.

11

Comets and Shooting-Stars

The observatory on Palomar mountain, in California, is one of the most famous in the world. Its main instrument is the 200-inch (508-cm) Hale reflector, which is used mainly for studying remote stars and star-systems, and which has been in action ever since 1948, revolutionizing many of our ideas about the universe. For many years it was the largest of all telescopes, and though this is no longer true it is still in the forefront of astronomical research.

On 16 October 1982 the Hale reflector was used for a different purpose. Equipped with an ultra-sensitive electronic device, it was turned toward the position in the sky where Halley's Comet was expected to appear. It was an exciting search, and it was successful; there was a tiny blur of light, almost as faint as anything ever recorded, indicating that the comet was coming back after an absence of three-quarters of a century.

The recovery of Halley's Comet caused a tremendous surge of interest about comets in general, even though it cannot be said that at this return the comet had any prospect of becoming spectacular (as it has often been in the past). But it is quite true to say that in cometary studies, amateurs can play a very useful rôle, even if equipped with nothing more than binoculars.

Comets have been termed the stray members of the Solar System. They are not solid, rocky bodies, and they are of very low mass; a comet with a head larger than the Sun will have only a tiny fraction of the mass of the Moon or even a major asteroid. In ancient times bright comets caused a great deal of alarm and despondency, and the fear of them is not quite dead even today (at the 1910 return of Halley's Comet, an enterprising American made quite a large sum of money by selling anti-comet pills, though I have never been able to find out what they were meant to do). A large comet is made up of a nucleus, surrounded by a head or coma, from which issues a tail (or tails). Smaller comets may be tailless, so that they look deceptively like faint star-clusters or nebulæ.

One point is worth making at the outset. A comet moves well beyond the Earth's atmosphere, at a distance of millions of kilometres. Therefore it does not shift perceptibly, and you will have to watch it for hours before you can detect any motion at all against the starry background. An object which shifts noticeably will be nearby: a meteor, an artificial satellite, or something more mundane such as a high-altitude aircraft or a weather balloon.

A cometary nucleus is made up of icy material, together with solid particles. When the comet nears the Sun, the ice begins to evaporate, and the nucleus is hidden behind a kind of screen. The head or coma consists of very tenuous gas together with 'dust', and may be very large indeed, appearing so brilliant that the mediæval fear of comets is understandable.

There are two main types of tails: those made of gas, and those made of dust. The gas-tail is usually straight, and is almost incredibly rarefied, with a density far less than that of what we normally call a laboratory vacuum. The dust-tail is curved, and is composed of very small particles. Gas- or ion-tails always point more or less away from the Sun, because the ion-tail is repelled by the so-called solar wind – that is to say, a stream of atomic particles being sent out by the Sun constantly in all directions, and the dust-tail is repelled by the pressure of sunlight. A comet which is moving in toward the Sun will travel head-first, but after it has passed perihelion and has started to move outward it will move tail-first. Tails develop only when the comet is sufficiently close to the Sun to be warmed. Inevitably material is lost, and many of the short-period comets have lost so much of their volatile material that they are no longer capable of developing tails at any time.

Comets are bona-fide members of the Solar System. This seems fairly certain; some astronomers still maintain that they come from outer space, and are captured by the Sun during our passage through a spiral arm of the Galaxy, but the evidence as it stands is decidedly against anything of the sort. It is now believed that the short-period comets form a cloud of icy objects (the Kuiper Cloud or Kuiper Disk) not too far beyond the orbit of Neptune. When one of these objects is perturbed for any reason, it moves inward towards the Sun; it may then be 'captured' by a planet, usually Jupiter, and forced into an orbit with a period of a few years – only 3.3 years in the case of Encke's Comet, 6.2 years for D'Arrest's Comet, and so on. Some comets may actually commit suicide by falling into the Sun, and in July 1994 Comet Shoemaker-Levy 9 impacted Jupiter, producing disturbances there which remained visible for many months. It is significant that all these short-period comets move in the same sense as the Earth (direct motion), and many of them have their aphelion points at about the distance of the orbit of Jupiter, making up what has come to be called Jupiter's comet family.

Other comets have periods of over 50 years, and some of these, including Halley's, have retrograde motion. These, and also the so-called 'great' comets, are thought to come from the much more remote Oort Cloud, at more than a light-year from the Sun: they are often called non-periodical, though this is not strictly correct. When an Oort Cloud object is perturbed for any reason it may take hundreds, thousands or even millions of years to reach the inner Solar System and come within our range. Unless it is captured, it may swing round the Sun and return to the Cloud, not to be back again for an immensely long period. Some comets may be so affected by planetary perturbations that they are thrown out of the Solar System altogether – as with Comet Arend–Roland of 1957, which was a bright naked-eye object for some weeks in April of that year.

Most of the planets have orbits which are practically circular. Not so with the comets, almost all of which move in highly eccentric path. Look for instance at D'Arrest's Comet, one of the brighter members of Jupiter's family which can come within binocular range and even reaches naked-eye visibility occasionally. At perihelion its distance from the Sun is 1.2 astronomical units or 180 000 000 km; at aphelion it moves out to 840 000 000 km (remember that one astronomical unit is the Earth–Sun distance). D'Arrest's Comet has a period of 6.2 years and, like Encke's, is a well-known and regular visitor.

On the other hand, consider Kohoutek's Comet, which caused a tremendous amount of interest in 1973 because it was expected to become brilliant – though in the event it failed to do so. Its orbit is so eccentric that it does not differ much from an open curve or parabola. It will certainly return, but not for over 70 000 years, so that we have a long wait before it comes back into view. Comets depend upon reflected sunlight, so that they are visible only when they are reasonably close to the Sun and the Earth. It is true that near perihelion their material is excited sufficiently to emit a certain amount of light on its own account, but without the Sun a comet does not shine at all.

Halley's Comet is in a class of its own. The period is 76 years, though this is not quite constant: since a comet is of very low mass, it is at the mercy of the gravitational pulls of the planets, and in Halley's case the period is variable by a few years either way. Records of it go back well before the time of Christ, and every return since that of BC 87 has been documented. Halley's is the only bright comet which can be predicted, which is why the latest return, that of 1986, caused so much interest; after all, this was the first time that rocket probes could be sent to it.

Most comets are named after their discoverers or co-discoverers which can lead to some tongue-twisters; for instance Schwassmann–Wachmann, Churyumov–Gerasimenko and Smirnova–Chernykh. Less frequently the name honours the mathematician who first worked out the orbit. This is the case with Halley's Comet. In 1682 a bright comet was observed by Edmond Halley, afterwards Astronomer Royal; he calculated that the path was almost the same as those of comets previously seen in 1607 and 1531, and predicted that the three comets were one and the same, so that the next return would be in 1758. Up to that time it had been thought that comets moved in straight lines rather than curves, so that Halley's suggestion was a bold one; but it was correct, and on Christmas Night 1758 the comet was duly recovered, though, sadly, Halley was long since dead. It was clearly appropriate to name the comet in his honour.

At some returns Halley's Comet has been magnificent. In 837 it passed within 7 000 000 km of the Earth, and showed a head as bright as Venus with a tail stretching right across the sky. It was seen in 1066, when Duke William was preparing to invade England; King Harold seems to have regarded it as an evil omen, and it is shown on the famous Bayeux Tapestry, said by some authorities to have been woven by the Conqueror's wife. In 1301 it was seen by the Florentine artist Giotto di Bondone who used it as a model for the Star of Bethlehem in his picture 'The Adoration of the Magi' (which is why the European Space Agency probe to the comet more than six hundred years later was named Giotto). In 1456 it was again bright, and the current Pope, Calixtus III, preached against it as an agent of the Devil. In 1835 and again in 1910 it made a brave showing, and in 1910 the Earth actually passed through its tail, though no effects could be detected. But everything depends upon the relative positions of the Earth, the Sun and the comet as Halley moves in toward perihelion, and the 1986 return must be classed as the least favourable for the past two thousand years. Things will be no better at the next return, that of 2061.

Because a comet loses some of its material, by evaporation. each time it passes through perihelion, there must come a time when all the volatiles are exhausted. In fact comets must be short-lived by cosmical standards, and there are several known cases of short-period comets which have now disintegrated. Biela's Comet, which used to have a period of 6.6 years, broke in two at the 1845 return, and has never been seen since 1852; there can be little doubt that it is defunct, and so is another last-century periodical comet, Brorsen's. Westphal's Comet, with a period of 62 years, faded out during the return of 1913 and failed to reappear. There was also the bright West's Comet of 1976, which showed obvious signs of breaking up after it had passed perihelion. It is not likely to be brilliant at its next return, but in any case it will not come back for thousands of years.

Some past comets have been spectacular enough to be seen in broad daylight. Such were the comets of 1811, 1843, 1858 and others. The last 'daylight comet' was that of January 1910, which was at its best in the first days of the year, a few weeks before Halley's. Since then there have been various naked-eye comets, notably Arend–Roland (1957), Bennett (1970) and West (1976), but two comets which appeared near the end of our own century almost qualified as 'great'. The first of these was that of March 1996, discovered by the Japanese amateur Hyakutake; it was in fact a small comet, but made a relatively close approach, and was really spectacular, with a long tail. At its best it was in the far north of the sky and so ideally placed for observers in Britain and the northern United States; one unusual characteristic was its beautiful green colour. Then, in the early months of 1997, came an even brighter comet, Hale–Bopp – discovered by two American amateurs. The comet developed a broad, curved and slightly reddish dust-tail and a long, bluish gas-tail; it became so bright that it could not possibly be overlooked. The orbital period is around 4000 years, while to see Hyakutake again we must be resigned for a wait of at least 15 000 years.

Binoculars are ideal for making observations of bright comets. If the tail is long, a low magnification and a wide field will be needed. I remember that I preferred 7 × 50 for the bright Bennett's Comet of 1970, though × 12 gave good views of the details in the tail, and these details are always liable to change quickly. For fainter comets, higher powers may be used, though of course it is often difficult to locate a dim comet with × 20 or so unless its position is very accurately known.

Quite apart from this, there is the favourite amateur pursuit of comet-hunting. Non-professional astronomers have a fine record, and will no doubt continue to make discoveries regularly.

A high-power telescope is of no use at all in searching for new comets. The essential is to have a wide field, adequate magnification, and a really good knowledge of the sky (remember Charles Messier!). Specialist binoculars are a great help. The most successful comet-hunters, such as George Alcock in England and many of the Japanese amateurs, have pairs equipped with lenses up to 150 millimetres in diameter; with a low magnification, these are ideal.

However, binoculars of this size are prohibitively expensive nowadays, and most hunters will have to make do with more conventional pairs. The only advice I can offer is to look for binoculars with as large an aperture as possible, with a magnification of no more than × 12 and preferably rather less.

The best regions in which to search are the western sky after sunset and the east before dawn, but since a new comet will probably be faint there is no point in observing unless the sky is fairly dark. Simply make a series of sweeps, from left to right (or vice versa), starting fairly high up and making each sweep progressively lower than the last, taking care to overlap. You will certainly pick up many clusters and nebulæ, so that it is essential to have a star-map handy. Also, beware of faint groups of stars which are not true clusters, and which are therefore not listed, but which can still be confused with comets. It is not really likely that an unexpected comet will have a tail long enough and bright enough to be seen in binoculars, though of course one never knows.

If you decide to go comet-hunting, be prepared for many hundreds of hours of fruitless sweeping before making a discovery; you may be lucky straight away, or you may never have any success. Yet finding a comet is truly exciting. I have experienced it once, though I was busy observing variable stars at the time and was not looking for comets. As I swept on to a variable star in Taurus, an unmistakable comet came into view. It was of about the eighth magnitude, and I was not naïve enough to believe that I was the first to see it; I have to be honest, and admit that it was none other than Encke's Comet, which had been under observation for months. All the same, it was a great moment.

When a brilliant comet appears, binoculars can give good views of the short-term changes in the tail and even the coma. Observations of this sort are very useful, because the changes can be surprisingly rapid, and it is important to keep them under constant watch.

As a comet moves along, it leaves a 'trail' of what may be called dust. When one of these particles dashes into the Earth's air, it burns away by friction to produce a meteor or shooting-star. There are various annual showers; each time we pass through a shoal of meteors, the Earth collects many shooting-stars. The most reliable shower is that of early August, associated with a comet (Swift–Tuttle) which has a period of 130 years and last returned in 1992. Other showers are those of December (the Geminids) and January (the Quadrantids, coming from a region of the sky near the star Beta Boötis which used to be included in the now-rejected constellation of Quadrans, the Quadrant). Two annual showers, those of May and October, are associated with Halley's Comet, while the Andromedids of November mark the remnants of the lost Biela's Comet. The Leonids, also of November, can be spectacular at times, as in 1966, though generally they are sparse.

We expected Leonid meteor storms in 1998 and 1999, and they did occur; that of 1998 was rather earlier than expected, and of brief duration – unfortunately most of Britain was clouded out. The 1999 meteor storm lasted for less than an hour on the early morning of 18 November, and again England was mainly cloudy, though Scotland was more favoured. After 2000 and perhaps 2001 we can expect no more spectacular Leonid storms for over thirty years. Apart from the showers, there are also sporadic meteors, which may appear from any direction at any moment.

The meteors of any particular shower travel through space in parallel paths, so that they seem to issue from a definite point or radiant in the sky: the Leonids from Leo, and so on. Meteor-plotting is a useful amateur pursuit, but I do not propose to say more about it here, simply because binoculars are of no use whatsoever. The best instrument is the naked eye.

En passant, artificial satellites can be tracked, because – unlike meteors – they move slowly enough to be followed with binoculars. In the early days of space research, amateurs played a major part by noting the moments when satellites passed close to known stars, so that their positions could be fixed accurately. This is still done, though we must admit that nowadays it is much less important than it used to be – because we know so much more about the density of the upper air, which regulates the ways in which the lower-altitude satellites move. It is surprising how often the satellites come into view with binoculars. Recently I was making an observation of a variable star, using 7 × 50, when over a period of five minutes I saw no less than four satellites crawl across the field of view.

Halley's Comet, 1985 November 17, photographed by Patrick Moore from Selsey. Exposure 10 minutes. 200-mm telephoto lens, guided: film, Fuji ASA 100.

12

Summing-Up

In this book I have done my best to outline what the binocular-user may expect to see in the sky. Let me end by summarizing the situation as concisely as I can.

1. **Binoculars** of all kinds may be used. A low magnification will give a wider field; a higher magnification will of course show fainter objects and more detail. For an all-purpose pair, 7 × 50 is ideal. Powers of from 7 to 12 can be hand-held without much difficulty, but with a higher magnification some sort of mounting is highly desirable, and for above × 15 or so it is almost essential.

2. **The Sun.** It is possible to project the solar image on to a screen suitably held or fixed behind the eyepieces. Sunspots may be seen in this way, but not very well, and frankly I would not recommend trying to observe the Sun without a telescope. Under no circumstances look direct, even by using a dark filter when the Sun is low down and looks deceptively harmless.

3. **The Moon.** There is no danger to the eye. Many lunar features can be seen, ranging from maria to mountains, valleys, craters and rills. Details are best seen near the terminator, and a little patience will soon enable you to find your way around. Use as high a magnification as is practicable.

4. **The Planets.** Mercury can be easily picked up with binoculars after sunset or before sunrise, when suitably placed. Do not sweep for it unless the Sun is completely below the horizon. With × 20 the phase may be glimpsed, but not at all easily.
   Venus shows its phases, easily detectable except when near full.
   Mars shows a small red disk; nothing more.
   Jupiter shows an obvious disk. Surface details will not be seen, but it should be possible to pick up the Galilean satellites with good binoculars; the higher the magnification, the easier it will be.
   Saturn's rings are out of binocular range, and so are all the satellites, except possibly Titan, but you will certainly see that there is something unusual about the planet's shape – provided that the ring system is wide open, as it will be in the first years following the end of the century.
   Uranus is easily visible, and × 12 or over may be enough to show that it is not a star.

Neptune and some of the asteroids can be found, but will look stellar, and the only way to identify them is by their movements from one night to another. Pluto is completely out of binocular range.

5. **Comets** are well seen in binoculars if bright, and details in the head and tails can be followed as they change, sometimes quite quickly. For this, a reasonably high power is best; say × 12. Comet-hunting is a pastime in which amateurs have been most successful; the need here is for a low magnification and as large an aperture as possible.

6. **Artificial satellites** may be tracked, and their positions fixed by timing them as they pass close to known stars. Meteors (and also the lovely sky glows such as auroræ and the Zodiacal Light) are not suited for binocular work.

7. **The Stars.** The colours are brought out well, particularly with red stars. Many doubles can be found; with a wide pair use a low power, though for closer pairs a higher magnification is needed. There are many variable stars within binocular range, and useful scientific work may be carried out by estimating their magnitudes against non-variable comparison stars. Nova-hunting can be carried out with binoculars. A wide field and as large an aperture as possible is needed, together with an encyclopædic knowledge of the sky. The best areas to search are those in and near the Milky Way, where most novæ appear.

8. **Star-clusters** in large numbers are available for inspection, both open and globular. With large open clusters, such as the Pleiades and Presepe, use a low power and a wide field. With fainter clusters, including the globulars, a higher magnification will give the best results.

9. **Nebulæ** and **Galaxies** are less common so far as the binocular-owner is concerned, but there are still a good number of them, and some, such as the Orion Nebula and the Magellanic Clouds, are spectacular. Again, the magnification used depends upon the size of the object, together with the surface brightness. Some galaxies, notably the Triangulum Spiral, are well seen with × 7, but are decidedly elusive in a small telescope. Those with a good knowledge of the sky can hunt for supernovæ in external galaxies. The chances of success may be slight, but certainly not nil.

Finally, I would strongly recommend joining an astronomical society – for instance, the British Astronomical Association. Most towns and cities in Britain, the United States, Australia, New Zealand and South Africa have their local societies.

So if you do not wish to acquire a telescope, or cannot afford to do so, do not be depressed. Equip yourself with binoculars, and I promise you that you will not be disappointed. There is always plenty to see.

I wish you clear skies – and the best of luck!

Appendix 1

Choosing a Telescope

Useful though binoculars are, it cannot be denied that they have marked limitations, the worst of which is the lack of sheer magnification. If you want to see the rings of Saturn, the polar caps of Mars, the Red Spot of Jupiter or the myriad stars of a globular cluster, you must have a telescope; and this is often where trouble begins, because a telescope is either good or cheap – not both.

To recapitulate: a very small telescope is of little use, and the best that can be said of it is that it is an improvement upon nothing at all. The minimum useful size is an aperture of 3 inches for a refractor, or 6 inches for a Newtonian reflector.* Remember that the maximum useful magnification is x 50 per inch of aperture, so that, for example, a 3-inch refractor will never stand more than x 150 – and then only if the optics are good.

The instinctive temptation is to rush into the nearest camera shop and buy something such as a 2- or 2½-inch Japanese refractor, or a 3- or 4-inch Newtonian. It will almost certainly look nice, and it will be an attractive ornament, but even though it will cost from £100 to £300 ($150–450) it is almost certain to be a grave disappointment. And beware of even smaller telescopes, with cardboard tubes, priced at around £30 to £50 ($40-75). I have one of these, and to be candid I can see a distant object rather better without it. Quite apart from anything else, these small telescopes will give inconveniently small fields of view.

There are some notorious traps. In particular, avoid any telescope which is advertised by its magnifying power; for example I have seen a recent advertisement claiming that the telescope concerned will 'give a magnification of 600 times' – but as the aperture is only 3 inches, this is absolutely absurd. Remember, magnification is due entirely to the eyepiece; it is the function of the object-glass or mirror to collect the light in the first place, and if there is not enough light available the resulting image will be hopelessly dim. Therefore, always check first on the telescope's aperture – that is to say, the diameter of the object-glass for a refractor or the main mirror for a reflector. Advertisers are depressingly fond of using this trick.

---

* Forgive me for using Imperial units; in this particular context the inch is more convenient than the centimetre. Thus 3 inches is equal to 7.6 centimetres; it is easy to convert if you wish to do so.

Next, look for an 'aperture stop' fitted into the tube either above or below the object-glass, thereby reducing the effective light-gathering area. This is done to cut out the worst part of a poor-quality lens (the trick is less straight-forward for a reflector, but it can still be played). I was recently shown a good-looking refractor which was said to have an aperture of 2½ inches. Cunningly fitted inside the tube, out of immediate sight, was a ring which cut the real aperture down to just over one inch.

Another problem is that a bad telescope does not always betray itself at first glance. Short of making a strict optical check, the only course is to try it out before purchasing, which is clearly much less easy with a telescope than with binoculars; there may be an unacceptable amount of false colour in a refractor, and it may prove impossible to obtain a sharp focus with either a refractor or a reflector. Another common defect is that the eyepiece rack will not move in or out sufficiently to give a sharp focus, particularly with a higher-powered eyepiece.

What, then, is a safe procedure?

Buying a telescope by mail order is always a risk; you may be lucky, you may not – and we have to admit that there are some quite well-known tele-scope firms which supply poor instruments. If you can, have your proposed telescope checked or at least recommended by an expert. Your local astro-nomical society can usually help here (there is a full list of societies in the annual *Yearbook of Astronomy*). Second-hand telescopes used to be easy to obtain, but today they are not. Of course, there is always a chance of a good 'buy', but my advice is never to purchase a second-hand telescope before trying it out or having it vetted by an expert.

All in all, you must be prepared for any outlay of at least £350 ($500), and probably rather more. This seems a considerable sum, but bear in mind that the cost is non-recurring, because a telescope will last you a lifetime pro-vided that you take proper care of it.

If finance is an insuperable problem, the only alternative is to make your own telescope. Frankly, lens-grinding is a task for the really knowledgeable and well-equipped expert, and to buy an object-glass and then mount it is not really sensible, because the main cost of a refractor is in the object-glass itself. On the other hand, making a mirror for a Newtonian is much less of a problem, and can be done by anyone who is good with his hands – though it is very time-consuming, and the beginner must be prepared for several fail-ures before turning out a really good mirror. This is no place to go into the details of mirror-making, even if I were competent to do so (which I am not), but there are various books available.

If you buy the blanks for a Newtonian, grind the main mirror and make the mounting, you can acquire a really good 6-inch at relatively low cost. If you do not want to risk mirror-grinding, you can always buy the mirror and then mount it, which is admittedly more expensive but is cheaper than buying the whole telescope ready for use. If you are going to try your hand at telescope-construction, I would most certainly suggest joining your local society.

The straightforward altazimuth mounting has its advantages for a small but adequate telescope; the Dobsonian, which can be made entirely of wood, is a simple variant of this. There is not a great deal to go wrong; the telescope can be swung quickly from one part of the sky to another without the need for complicated manœuvres, and neither is there any need to spend a long time in adjusting the polar axis. (There is nothing more infuriating than a badly-aligned equatorial.) Against this, the telescope must be moved constantly by hand to follow the target object as it tracks across the sky, and two slow motions are needed, one for altitude and the other for azimuth, so that if notes are to be taken at the same time the observer ideally needs three hands. If you mean to use the telescope for photography, you must have an equatorial, preferably mechanically driven. Of course 'folded' or catadioptric telescopes such as Schmidt–Cassegrains are ideal, and have the added advantage of being portable, but the cost is very much higher, so that I will say no more about them here.

Many people will not want to go to these lengths. In that case, I hope I have shown that a tremendous amount of enjoyment can be had from equipping yourself with binoculars. The skies are always changing, and there is always something new to see.

Appendix 2

Planetary Data

Planet	Mean distance from Sun, millions of km	Orbital period	Axial rotation period	Axial inclination, degrees	Equatorial diameter, km	Max. magnitude
Mercury	58	88 d	58.6 d	2	4878	−1.9
Venus	108	224.7 d	243.2 d	178	12 104	−4.4
Earth	150	365.3 d	23h 56m	23.5	12 756	−
Mars	228	687 d	24h 37m	24	6 794	−2.8
Jupiter	778	11.9 y	9h 50m	3	143 884	−2.6
Saturn	1427	29.5 y	10h 14m	26	120 536	−0.3
Uranus	3004	84.0 y	17h 14m	98	51 118	+5.6
Neptune	4537	164.8 y	16h 6m	29	50 538	+7.7
Pluto	5900	247.7 y	6d 9h	122	2324	+13.9

Planetary satellites over 3000 kilometres in diameter

Satellite	Primary	Mean distance from primary, centre to centre (km)	Orbital period, days	Diameter, km	Mean opposition magnitude
Ganymede	Jupiter	1 070 000	7.15	5268	4.6
Titan	Saturn	1 221 860	15.94	5150	8.4
Callisto	Jupiter	1 880 000	16.69	4806	5.6
Io	Jupiter	421 600	1.77	3660	5.0
Moon	Earth	384 400	27.32	3476	−12.7
Europa	Jupiter	670 900	3.55	3130	5.3

Appendix 3

The Planets, 2000–2010

Mercury

Western elongations:	2000 Mar. 28, July 27, Nov. 15.
	2001 Mar. 11, July 9, Oct. 29.
	2002 Feb. 21, June 21, Oct. 13.
	2003 Feb. 4, June 3, Sept. 27.
	2004 Jan. 17, May 14, Sept. 9, Dec. 29.
	2005 Apr. 26, Aug. 23, Dec. 12.
	2006 Apr. 8, Aug. 7, Nov. 25.
	2007 Mar. 22, July 20, Nov. 8.
	2008 Mar. 3, July 1, Oct. 22.
	2009 Feb. 14, June 13, Oct. 6.
	2010 Jan. 27, May 26, Sept. 20.
Eastern elongations:	2000 Feb. 15, June 9, Oct. 6.
	2001 Jan. 28, May 22, Sept. 18.
	2002 Jan. 11, May 4, Sept. 1, Dec. 26.
	2003 Apr. 16, Aug. 14, Dec. 9.
	2004 Mar. 29, July 27, Nov. 21.
	2005 Mar. 12, July 9, Nov. 3.
	2006 Feb. 24, June 20, Oct. 17.
	2007 Feb. 7, June 2, Sept. 29.
	2008 Jan. 22. May 14, Sept. 11.
	2009 Jan. 4, Apr. 26, Aug. 25, Dec. 18.
	2010 Apr. 9, Aug. 6, Dec. 2.

Venus

Western elongations: 2001 June 8, 2003 Jan. 11, 2004 Aug. 17, 2006 Mar. 25, 2007 Oct. 28, 2009 June 7.

Eastern elongations: 2001 Jan. 17, 2002 Aug. 22, 2004 Mar. 29, 2005 Nov. 3, 2007 June 9, 2009 Jan. 14, 2010 Aug. 20.

Mars

Opposition date	Max. apparent diameter (")	Max. magnitude	Constellation
2001 June 13	20.8	−2.1	Sagittarius
2003 Aug. 28	25.1	−2.7	Capricornus
2005 Nov. 7	20.2	−2.1	Aries
2007 Dec. 24	16.3	−1.4	Gemini

Jupiter

Opposition date	Max. magnitude
2000 Nov. 28	−2.4
2002 Jan. 1	−2.3
2003 Feb. 2	−2.1
2004 Mar. 4	−2.0
2005 Apr. 3	−2.0
2006 May 4	−2.0
2007 June 5	−2.1
2008 July 9	−2.3
2009 Aug. 15	−2.1
2010 Sept. 21	−2.1

Saturn

Opposition date	Max. magnitude
2000 Nov. 19	−0.1
2001 Dec. 3	−0.3
2002 Dec. 17	−0.3
2003 Dec. 31	−0.3
2005 Jan. 13	−0.2
2006 Jan. 27	0.0
2007 Feb. 10	+0.2
2008 Feb. 24	+0.4
2009 Mar. 8	+0.4
2010 Mar. 22	+0.4

Appendix 4

Eclipses, 2000–2010

Solar eclipses

Date	Time (GMT)	Type	Maximum duration (if total or annular) m s	Percent eclipsed (if partial)	Area
2000 Feb. 5	13	Partial	–	56	Antarctic
2000 July 31	02	Partial	–	60	Arctic
2000 Dec. 25	18	Partial	–	72	Arctic
2001 June 21	12	Total	4 56	–	Atlantic, S. Africa
2001 Dec. 14	21	Annular	3 54	–	C. America, Pacific
2002 June 10	24	Annular	1 13	–	Pacific
2002 Dec. 4	08	Total	2 04	–	S. Africa, Australia
2003 May 31	04	Annular	3 37	–	Iceland
2003 Nov. 23	23	Total	1 57	–	Antartica
2004 Apr. 19	14	Partial	–	74	Arctic
2005 Apr. 8	21	Total	0 42	–	Pacific, N. of S. Amer.
2005 Oct. 3	11	Annular	4 32	–	Spain, Africa
2006 Mar. 29	10	Total	4 07	–	Atlantic, Turkey, Russia
2006 Sept. 22	12	Annular	7 09	–	Atlantic, S. Indian Ocean
2007 Mar. 19	03	Partial	–	88	
2007 Sept. 11	13	Partial	–	75	
2008 Feb. 7	04	Annular	2 14	–	S. Pacific, Antarctica
2008 Aug. 1	10	Total	2 28	–	N. Canada, Arctic, Siberia, China
2009 Jan. 26	05	Annular	7 53	–	Indian Ocean

Date	Time (GMT)	Type	Maximum duration (if total or annular) m s	Percent eclipsed (if partial)	Area
2009 July 22	03	Total	6 40	–	W. Pacific
2010 Jan. 15	05	Annular	11 07	–	C. Africa, Indian Ocean
2010 July 11	20	Total	5 20	–	S. Pacific

Lunar eclipses

Date	Type	Percent eclipsed (if partial)	Time of mid-eclipse, GMT h m	Duration of eclipse Totality h m	Partial h m
2000 Jan. 21	Total	–	4 45	1 16	3 22
2000 July 16	Total	–	13 57	1 00	3 16
2001 Jan. 9	Total	–	20 22	0 30	1 38
2001 July 5	Partial	49	14 57	–	1 19
2003 May 16	Total	–	03 41	0 26	1 37
2003 Nov. 9	Total	–	01 20	0 11	1 45
2004 May 5	Total	–	20 32	0 38	1 41
2004 Oct. 28	Total	–	03 95	0 40	1 49
2005 Oct. 17	Partial	06	12 04	–	0 28
2006 Sept. 7	Partial	18	18 52	–	0 45
2007 Mar. 3	Total	–	23 22	0 37	1 50
2007 Aug. 28	Total	–	10 38	0 45	1 46
2008 Feb. 21	Total	–	03 27	0 24	1 42
2008 Aug. 16	Partial	81	21 11	–	1 34
2010 Dec. 21	Total	–	8 18	1 12	

There will be penumbral eclipses on 2001 Dec. 30 (89), 2002 May 6 (69), 2002 June 24 (21), 2002 Nov. 20 (86), 2005 Apr. 24 (87), 2006 Mar. 14 (100).

Appendix 5

The Brightest Stars

The angular distance of a star north or south of the equator is its declination (the celestial equivalent of latitude). To find out whether or not a star can be seen at any time, simply subtract your own latitude on the Earth's surface from 90; this gives your co-latitude. Thus if your latitude is 52 degrees north (about that of London) the co-latitude is 90 – 52 = 38. This means that any star whose declination is north of +38 degrees will be circumpolar (i.e. will never set) and any star south of declination –38° will never rise. Thus Deneb (dec. +45°) is circumpolar from London, but Canopus (dec. –53°) can never be seen. The declination of Acrux, the brightest star in the Southern Cross, is –63°, so that it rises from anywhere on the Earth's surface south of latitude 27 degrees north.

The following are the brightest stars in the sky, officially ranked as being of the first magnitude:

Star	Name	Magnitude	Spectrum	Declination, degrees
α Canis Majoris	Sirius	–1.5	A1	–16
α Carinæ	Canopus	–0.7	F0	–53
α Centauri	–	–0.3	K1+G2	–61
α Boötis	Arcturus	–0.0	K2	+19
α Lyræ	Vega	0.0	A0	+39
α Aurigæ	Capella	0.1	G8	+46
β Orionis	Rigel	0.1	B8	–08
α Canis Minoris	Procyon	0.4	F5	+05
α Eridani	Achernar	0.5	B5	–57
α Orionis	Betelgeux	var.	M2	+07
β Centauri	Agena	0.6	B1	–60
α Aquilæ	Altair	0.8	A7	+08
α Crucis	Acrux	0.8	B1+B3	–63
α Tauri	Aldebaran	0.8	K5	+17

THE BRIGHTEST STARS, CONTINUED

Star	Name	Magnitude	Spectrum	Declination, degrees
α Scorpii	Antares	1.0	M1	−26
α Virginis	Spica	1.0	B1	−11
β Geminorum	Pollux	1.1	K0	+28
α Piscis Aust.	Fomalhaut	1.2	A3	−30
α Cygni	Deneb	1.2	A2	+45
β Crucis	–	1.2	B0	−60
α Leonis	Regulus	1.3	B7	+12

Next in order, not officially ranked as of the first magnitude, come ε Canis Majoris (Adhara), α Geminorum (Castor), γ Crucis, λ Scorpii (Shaula), γ Orionis (Bellatrix) and β Tauri (Alnath), all of magnitude 1.6.

Appendix 6

A Selection of Stellar Objects

All these objects are shown in the maps, and the list given here is bound to reflect my personal preference, but it may be useful as quick reference.

Red and Orange Stars

Most of these are of types K or M. Their colours are well brought out with binoculars.

	Star	Magnitude		Star	Magnitude
β	Andromedæ	2.1	α	Hydræ	2.0
γ	Aquilæ	2.7	γ	Hydri	3.2
ε	Aræ	4.1	σ	Libræ	3.3
α	Boötis	−0.0	R	Leporis	5.5 max
ε	Boötis	2.4	α	Lyncis	3.1
W	Boötis	4.7 max	R	Lyræ	3.9 max
σ	Canis Majoris	3.5	δ	Ophiuchi	2.7
γ	Canis Minoris	4.3	β	Pegasi	2.3 max
ε	Carinæ	1.9	ρ	Persei	3.2 max
R	Carinæ	3.8 max	L²	Puppis	2.6 max
μ	Cephei	3.4 max	α	Orionis	0.1 max
α	Ceti	2.5	η	Sagittarii	3.1
o	Ceti	1.7 max	α	Scorpii	1.0
γ	Crucis	1.6	κ	Serpentis	4.1
λ	Draconis	3.8	α	Trianguli Aust.	1.9
γ	Eridani	2.9	μ	Ursæ Majoris	3.0
η	Geminorum	3.1 max	β	Ursæ Minoris	2.1
μ	Geminorum	2.9	λ	Velorum	2.2
β	Gruis	2.1	N	Velorum	3.1
α	Herculis	3.1 max	δ	Virginis	3.4

Double Stars

Star	Mags.	Position angle, °	Separation, sec. of arc
μ Boötis	4.3, 7.0	171	109
α Capricorni	3.6, 4.2	291	378
β Capricorni	3.1, 6.0	267	205
β Cygni	3.1, 5.1	054	34
ν Draconis	4.9, 4.9	312	62
ε Lyræ	4.7, 5.1	173	208
ν Scorpii	4.3, 6.4	337	41
θ Serpentis	4.5, 4.5	104	22
κ -17 Tauri	4.2, 5.3	173	339
θ Tauri	3.4, 3.8	346	337
ζ-80 Ursæ Maj.	2.3, 4.0	071	709
α-8 Vulpeculæ	4.4, 5.8	028	414

Look also for the wide pairings α–θ Chamæleontis, γ–6 Equulei, o¹–o² Eridani, δ Gruis, μ Gruis, α Libræ, φ Lupi, δ¹–δ² Lyræ, λ–μ Musdæ, β¹–β² Sagittarii, ζ¹–ζ² Scorpii, μ Scorpii, δ–64 Tauri, β¹–β² Tucanæ.

Variable Stars

Star	Type	Range	Period, days
R Andromedæ	Mira	5.8–14.9	409
η Aquilæ	Cepheid	3.5–4.4	7.2
ε Aurigæ	Eclipsing	3.0–3.9	9892
ε Aurigæ	Eclipsing	3.7–4.1	972
W Boötis	Semi-regular	4.7–5.4	±450
X Cancri	Semi-regular	5.6–7.5	±195
UW Canis Majoris	β Lyræ	4.0–5.3	4.4
η Carinæ	Irregular	−0.8–7.9	–
ZZ Carinæ	Cepheid	3.3–4.2	36
U Carinæ	Cepheid	5.7–7.0	39
R Carinæ	Mira	3.9–10.5	309
α Cassiopeiæ	Suspected	2.0–2.4	–
γ Cassiopeiæ	Irregular	1.6–3.3	–
ρ Cassiopeiæ	?	4.1–6.2	–

Variable Stars, continued

Star	Type	Range	Period, days
R Centauri	Mira	5.3–11.8	546
δ Cephei	Cepheid	3.3–4.4	5.4
μ Cephei	Semi-regular?	3.4–5.1	?
W Cephei	Eclipsing	4.8–5.4	7430
o Ceti	Mira	1.7–10.1	332
R Coronæ Borealis	Irregular	5.7–15	–
T Coronæ Borealis	Recurrent nova	2.0–10.8	–
χ Cygni	Mira	3.3–14.2	407
P Cygni	P Cygni	3–6	–
U Cygni	Mira	5.9–12.1	462
R Cygni	Mira	6.1–14.2	426
W Cygni	Semi-regular	5.0–7.6	±130
U Delphini	Semi-regular	5.6–8.0	±110
EU Delphini	Semi-regular	5.8–6.9	±59
β Doradûs	Cepheid	3.7–4.1	9.8
η Geminorum	Semi-regular	3.2–3.9	±233
ζ Geminorum	Cepheid	3.7–4.1	10.1
π¹ Gruis	Semi-regular	5.8–6.4	±150
α Herculis	Semi-regular	3-4	±100?
R Horalogii	Mira	4.7–14.3	404
R Hydræ	Mira	4.0–10.0	390
R Leonis	Mira	4.4–11.3	432
δ Librae	Algol	4.9–5.9	2.3
β Lyrae	β Lyræ	3.3–4.3	12.9
R Lyrae	Semi-regular	3.9–5.0	±46
χ Ophiuchi	Irregular	4.2–5.0	–
α Orionis	Semi-regular	0.1–0.8	±212
U Orionis	Mira	4.8–12.6	372
W Orionis	Semi-regular	5.9–7.7	±212
κ Pavonis	Type II Cepheid	3.9–4.7	9.1
λ Pavonis	Irregular	3.4–4.3	–
β Pegasi	Semi-regular	2.4–2.9	±38

Variable Stars, continued

Star	Type	Range	Period, days
β Persei	Algol	2.1–3.3	2.7
ρ Persei	Semi-regular	3.2–4.2	±45
X Persei	Irregular	6.0–6.6	–
ζ Phœnicis	Algol	3.6–4.1	1.7
TX Piscium	Irregular	6.9–7.7	–
L² Puppis	Semi-regular	2.6–6.2	±140
U Sagittæ	Algol	6.4–9.0	3.4
R Scuti	RV Tauri	4.4–8.2	±140
R Serpentis	Mira	5.1–14.4	356
λ Tauri	Algol	3.3–3.8	3.9
Z Ursæ Majoris	Semi-regular	6.8–9.1	±196

Clusters and Nebulæ

Object	Constellation	Type	Mag.	Diameter, min of arc
M31	Andromeda	Galaxy	3.5	178 × 63
M2	Aquarius	Globular	6.5	13
C63, NGC 7293 (Helix)	Aquarius	Planetary nebula	6.5	13
C86, NGC 6397	Ara	Globular	5.6	26
M36	Auriga	Open cluster	6.0	12
M37	Auriga	Open cluster	5.6	24
M38	Auriga	Open cluster	6.4	21
M44 (Præsepe)	Cancer	Open cluster	3.1	95
M67	Cancer	Open cluster	6.9	30
M3	Canes Venat.	Globular	6.4	16
M41	Canis Major	Open cluster	4.5	38
M30	Capricornus	Globular	7.5	11
C102, IC 2602 (θ Carinæ)	Carina	Open cluster	1.9	50
M52	Cassiopeia	Open cluster	6.9	13
C13, NGC 457 (φ Cas)	Cassiopeia	Open cluster	6.4	13
C80, NGC 5139 (ω Cen)	Centaurus	Globular	3.6	36
Mel 111	Coma Beren.	Open cluster	4	275

Clusters and Nebulæ, continued

Object	Constellation	Type	Mag.	Diameter, min of arc
NGC 6541	Corona Aust.	Globular	6.6	13
C94, NGC 4755 (Jewel Box)	Crux	Open cluster	4.2	10
C99, Coal Sack	Crux	Dark nebula	–	400 × 300
C20, NGC 7000 (N. America)	Cygnus	Nebula	6	120 × 100
M39	Cygnus	Open cluster	4.6	32
C103, NGC 2070 (Tarantula)	Dorado	Nebula in LMC.	4	40 × 25
M35	Gemini	Open cluster	5.0	28
M13	Hercules	Globular	5.9	17
M92	Hercules	Globular	6.5	11
M48	Hydra	Open cluster	5.8	54
M65	Leo	Galaxy	9.3	10 × 35
M66	Leo	Galaxy	9.0	9 × 4
M56	Lyra	Globular	8.2	7
C50, NGC 2244	Monoceros	Open cluster	4.8	24
M50	Monoceros	Open cluster	5.9	16
C89, NGC 6087(S Normæ)	Norma	Open cluster	5.4	12
M42 (Sword of Orion)	Orion	Nebula	5	66 × 60
M15	Pegasus	Globular	6.3	12
C14, NGC 869/884 (Sword-Handle)	Perseus	Open clusters	4.3, 4.4	30, 30
M34	Perseus	Open cluster	5.2	35
M47	Puppis	Open cluster	4.4	30
M23	Sagittarius	Open cluster	5.5	27
M21	Sagittarius	Open cluster	5.9	13
M25 (v Sagittarii)	Sagittarius	Open cluster	4.6	32
M28	Sagittarius	Globular	6.9	11
M22	Sagittarius	Globular	5.1	24
M20 (Trifid Neb)	Sagittarius	Nebula	7.6	29 × 27
M9 (Lagoon Neb)	Sagittarius	Nebula	6.0	90 × 40
M17 (Omega Neb)	Sagittarius	Nebula	7.0	46 × 37
M6 (Butterfly)	Scorpius	Open cluster	4.2	15
M7	Scorpius	Open cluster	3.3	80

CLUSTERS AND NEBULÆ, CONTINUED

Object	Constellation	Type	Mag.	Diameter, min of arc
M4	Scorpius	Globular	5.9	26
M80	Scorpius	Globular	7.2	9
M/11 (Wild Duck)	Scutum	Open cluster	5.8	14
M5	Serpens	Globular	5.8	17
M16 (Eagle neb)	Serpens	Nebula	6.4	35×28
M45 (Plelades)	Taurus	Open cluster	1.2	110
C41, Hyades	Taurus	Open cluster	1	330
M/1 (Crab Nebula)	Taurus	Supernova remnant	8.4	6×4
M33 (Pinwheel)	Triangulum	Galaxy	5.7	62×39
NGC 6025	Triangulum A.	Open cluster	5.1	12
Small Magellanic Cloud	Tucana	Galaxy	2.3	280×160
C106, NGC 104 (47 Tuc.)	Tucana	Globular	4.0	31
M81	Ursa Major	Galaxy	6.9	27×14
C85, NGC 2391 (o Velorum)	Vela	Open cluster	2.5	50
M27(Dumbbell)	Vulpecula	Planetary nebula	7.6	6×15

Bibliography

ARNOLD, H.J.P. *Astrophotography: An introduction.* Reed, 1993.

ARNOLD, H.J.P., DOHERTY, P., AND MOORE, P. *The Photographic Atlas of the Stars.* Institute of Physics Publishing, 1999.

EDBERG, S., AND LEVY, P. *Observing Comets, Asteroids, Meteors and the Zodiacal Light.* Cambridge, 1994.

FISCHER, D., AND DUERBECK, H. *Hubble: A new Window on the Universe.* Copernicus, 1995.

HOSKIN, M. (ed.) *Illustrated History of Astronomy.* Cambridge, 1997.

HOWARD, N. *Standard Handbook of Telescope Making.* Harper & Row, 1984.

JONES, K.G. *Messier's Clusters and Nebulae.* Cambridge, 1993.

KAUFMAN, W. *Discovering the Universe.* Freeman, 1994.

MALIN, D. *A View of the Universe.* Cambridge, 1993.

MITTON, J. *Penguin Dictionary of Astronomy.* Penguin, 1992.

MOBBERLEY, M. *Astronomical Equipment for Amateurs.* Springer Verlag, 1998.

MOORE, P. *Stargazing.* Cambridge, 2000.

MOORE, P. *Atlas of the Universe.* Phillips, 2000.

MOORE, P. *Eyes on the Universe.* Springer Verlag, 1997.

MOORE, P. *Patrick Moore on Mars.* Cassell, 1999.

MURDIN, P. *End in Fire.* Cambridge, 1990.

NICOLSON, I., *Unfolding Our Universe.* Cambridge, 2000.

NORTON, A.P. *Norton's Star Atlas.* Gall & Inglis, 1989.*

SCHAAF, F. *Comet of the Century.* Copernicus, 1997.

* This is to be preferred to the more recent *Norton 2000*, where the maps cannot easily be used for plotting and the handbook has become too bulky.

The *Yearbook of Astronomy* is published annually by Macdonald.

Monthly magazines: *Sky and Telescope* (Sky Publishing Corporation, USA), *Spaceflight* (British Interplanetary Society, UK), *Astronomy and Space* (Box 2888, Dublin, Ireland).

There are many national and local societies open to amateurs. The British Astronomical Association (Burlington House, London W1V 0NL) issues a bimonthly journal, and holds monthly meetings in London as well as meetings in other parts of Britain. In the US, the Astronomical Society of the Pacific and the American Association of Variable Star Observers issue their own regular journals. In Canada, the Royal Astronomical Society of Canada is open to amateurs, as is the Royal Astronomical Society of New Zealand.

A Note on Metrication

Many people still prefer Imperial units to Metric. It is easy enough to convert the one to the other, as follows:

To turn inches into millimetres, multiply by 25.4.
To turn inches into centimetres, multiply by 2.54.
To turn feet into metres, multiply by 0.305.
To turn miles into kilometres, multiply by 1.609.
To turn millimetres into inches, multiply by 0.039.
To turn centimetres into inches, multiply by 0.394.
To turn metres into feet, multiply by 3.281.

Temperatures. If F = degrees Fahrenheit and C = degrees Celsius:

$$F = (C \times 1.8) + 32$$

$$C = (F - 32) \div 1.8.$$

Index